ADVANCE PRAISE FOR

OIL CRISIS IN OUR OCEANS
by Barbara E. Ornitz, J.D.

"We have to control oil spills every day. Smaller 'everyday' spills can be as environmentally damaging as the less frequent 'big ones' – especially on coral reefs. In major oil-producing and transporting regions, such as the Red Sea, coral reefs have experienced severe and long-term decline due to chronic oil releases. Even under the best conditions, oil spill response is only marginally effective, once the oil is out of the vessel. The U.S. must stay the course on OPA 90 because *prevention* is the best way to control oil spills. *Oil Crisis In Our Oceans* presents a concise, useful and thoughtful overview of where the U.S. is headed with oil spill response and prevention, and provides a persuasive argument that 'oil has to travel first class.'"

Jacqueline Michel, Ph.D.
Director, Environmental Technology Division
Research Planning, Inc.

"It is very valuable to have this kind of third-party examination of the intricacies of the NRDA world."

James F. Bennett
Environmental Protection Specialist
U.S. Dept. of the Interior

"Caribbean coral reefs now have less living coral, more algae, and fewer lobsters, conchs and fishes than 10 to 20 years ago. Their degradation is the result of a combination of many factors, including storms, sedimentation, nutrification, boat groundings and oil pollution. This comprehensive book compels all of us to take responsibility for and create awareness of the decline of these diverse and beautiful ecosystems. Oil pollution is a reality, and oil spills will continue to occur in the Caribbean. This book is a dynamic reference on the value of coral reefs as well as the cause and consequences of human carelessness."

Caroline Rogers
Biologist and Chief Scientist
National Biological Survey
Virgin Islands National Park

OIL CRISIS IN OUR OCEANS

Coral: Roadkill on the Petrohighway

Barbara E. Ornitz, J.D.

Tageh Press
Glenwood Springs, Colorado

Tageh Press
P.O. Box 401
Glenwood Springs, CO 81602

Graphic Design and Production: Alyssa Ohnmacht Advertising & Design,
     Carbondale, CO
Graphic Recreation and Production: Lori Rattan
Cover Illustration: Siddhia Hutchinson, Vieques, PR
Printing: Publishers Press, Salt Lake City, UT
Back Cover Photo: PA1 Alastair Worden, USCG; reprinted with permission of
     U.S. Coast Guard MSO San Juan, PR
Author Photo: Chaz Evans, Crystal River Photography, Carbondale, CO

Library of Congress Cataloging-in-Publication Data

Ornitz, Barbara E.
    Oil crisis in our oceans – coral: roadkill on the petrohighway/ Barbara E. Ornitz.
    Includes bibliographic references, appendices and index.
    ISBN 0-9638385-1-2
    1. Oil Spills – United States. 2. Coral Reefs – International. 3. Marine Biology – International. 4.
Environmental Protection – U.S. 5. Law: Environmental Protection – U.S. and International. 6. Oceans:
Oil Spill & Pollution – U.S. 7. Pollution: Ocean – International. 8. Pollution: Oil – U.S.

363.738
074
1/09

"The only perfect oil spill response is prevention."

Capt. Robert G. Ross
U.S. Coast Guard

"Around the world coral reefs have suffered a dramatic decline in recent years. About 10 percent may already have degraded beyond recovery. Another 30 percent are likely to decline seriously within the next 20 years. It has been predicted that more than two-thirds of the world's coral reefs may collapse ecologically within the lifetime of our grandchildren, unless we implement effective management of these resources as an urgent priority."

Peter Hulm and John Pernetta
"Reefs at Risk – Coral reefs, human use and global change"
World Conservation Union, October 1993

"Oil has to go first class, not economy class, and not business."

Rear Admiral James Card
U.S. Coast Guard
1995 International Oil Spill Conference

ACKNOWLEDGMENTS

The Crew: Harlan Feder – editor and supporter in times of need; Lori Rattan, Alyssa Ohnmacht, Dana Knipe.

The Cover: Siddhia Hutchinson, artist, reef lover and friend.

Reviewers: James Bennett, Linda Burlington, Gilberto Cintron-Molero, Dr. Roy Darville, Dr. Fernando Gonzalez, Jacqueline Michel, Dr. Mark Miller, Mary Morton, Dr. Abimael Rodríguez, Caroline Rogers, Capt. Robert Ross, Dr. Carlos Santiago, Cheryl Scannel, CDR Edwin M. Stanton.

Contributors/Photographs/Diagrams: Peter Albers, Jose Berrios, Christopher Bordeaux, Crystal River Photography, Greg DeMarco, Tom Eason, Ginger Garrison, Great Barrier Reef Marine Park Authority, Paul Hankins, Marion King, Marine Safety Office, San Juan, Puerto Rico, Dr. Mark Miller, Mark Miller of NRC., Katherine Orr, Capt. Robert Ross, CDR E.M. Stanton, Jerry Snyder, Jamie Storrie, James H. Timber, Sen. Freddie Valentin, World Wildlife Fund.

Words of Wisdom: Matthew Garamone, Capt. Robert Ross, CDR E.M. Stanton.

Interview Subjects: Almost 100, for their patience, concern, information, and willingness to educate and provide information to others about caring for the environment.

Legislation: Bob Roach, Spiegel & McDiarmid, Washington, D.C. – who led me through the intricacies of the ever-changing D.C. political realm.

Office Staff: Sue Gardner, Vicki Vera-NeCamp.

Support: Family, Friends, Caddo Lake Institute, Shellman & Ornitz, P.C.

Don Henley – Mentor, guiding the way for many of us about protecting our world.

and

Dwight K. Shellman, Jr. – For showing me the true meaning and power of teamwork.

DEDICATION

To Dwight, and to all the responders of the *Morris J. Berman* oil spill, San Juan, Puerto Rico, January-April, 1994, for their tireless efforts to preserve, protect, clean and restore the marine environment damaged by the spill. Their achievement underscores more than ever why we must strive for greater prevention so such efforts will not have to be repeated.

CONTENTS

Introduction

SECTION III:
BERMAN, THE CATALYST:
LESSONS FOR FUTURE PREVENTION

INTRODUCTION

In the early morning hours of January 7, 1994, the tank barge *Morris J. Berman* ran aground on a reef off San Juan, Puerto Rico, spilling 798,000 gallons of No. 6 diesel fuel into Caribbean tropical waters. The accident was the result of a series of preventable man-made mistakes and poor judgment. Oil discharged across miles of coastline, fouling tourist beaches, historic fortresses, public property and marine wildlife habitats.

Why did *Berman* happen? How might the spill have been averted? And how can future *Bermans* be prevented? This book is the result of a three-year quest for answers to those questions – and more. How do oil spills affect marine ecosystems, particularly coral reefs, which are the biologically rich and diverse rainforests of our oceans? What effort is required to clean up a spill of this magnitude? What does spill response cost – and who pays?

The 1989 *Exxon Valdez* spill was a rude wake up call. We saw how our insatiable demand for oil can devastate a pristine environment. We can no longer treat our world – three-quarters of which is ocean – as a limitless dumping ground. "Out of sight, out of mind" begins to mean "out of existence."

Following *Exxon Valdez*, the shipping community took bolder steps to exert control over the transport of petroleum and petroleum products along petrohighways – those increasingly busy sea lanes and vulnerable choke points over which oil travels by tanker and tank barge. The United States led the way with OPA 90, a regulation best summed up as *the polluter pays*.

New questions arise. Is the U.S. willing to maintain its role as a leader in oil spill prevention? Will natural coral reefs disappear within our grandchildren's lifetimes? What can people do to become part of a meaningful solution? The oil crisis in our oceans is ultimately the result of our energy consumption demands.

This volume is divided into three sections that closely relate to each other, but which also stand independently to address answers to these questions. Section I provides a comprehensive case study of the *Morris J. Berman* spill and response. Section II examines the impact of oil on coral reef ecosystems, such as the one devastated by the *Berman* tank barge grounding. Section III offers a concise analysis of current U.S. and international laws, regulations and people-oriented actions taken to prevent and respond to oil spills.

The ultimate intention of this book is that the lessons learned from the *Berman* spill and response will be remembered so we will not be doomed to repeat them. The next $87 million we save by not having to clean up another *Berman* can be used in so many more productive ways to benefit mankind and the creatures who inhabit this planet we share.

– *Barbara E. Ornitz*, Aspen, CO and Vieques, PR

SECTION I

THE *BERMAN* SPILL AND RESPONSE

ANATOMY OF AN OIL SPILL:

How Did *Berman* Happen

JANUARY 1994: OIL SPILLS OFF PUERTO RICO

The *Morris J. Berman*, an oil tank barge the size of a football field loaded with 1.5 million gallons of heavy fuel oil, left San Juan, Puerto Rico on a routine two-day voyage to St. John, Antigua, pulled by the tugboat *Emily S*. A few hours after its departure, the tug's improperly repaired towline parted – for the *second* time that night – and the immense, powerless barge drifted west with the currents for several hours until it ran aground on a 7,000-year-old reef 200 yards off Punta Escambrón, in the heart of San Juan, Puerto Rico's tourist area, at approximately 4 a.m. January 7, 1994.

The barge scraped across the top of the reef, tearing gaping holes in its cargo tanks. Oil flooded out into predawn aquamarine Caribbean waters. The reef suffered partial destruction in some parts, and almost total destruction in others where it was reduced to rubble when the barge ground against it and came to a precarious rest.

The crew aboard the *Emily S* had failed to notice the loss of the barge, although they were well aware of the possibility that the *Berman* might disconnect from the tug. Earlier that evening, at about 12:30 a.m., the tow wire connecting the barge and tug broke for the first time, with the vessels only two miles from shore. A deck hand and the tug captain, Roy S. McMichael, made a temporary repair of the steel towline, hooked the oil barge and tug back together, and proceeded at full speed toward their destination.

Two crew members asleep on the barge at the time of the grounding awoke in alarm and reported the accident to the U.S. Coast Guard (USCG). Their request for immediate help set a massive response effort into motion. Coast Guard and contractor personnel began mobilizing and quickly blocked the entrance to the San Antonio Channel, the ecologically sensitive "back door" entrance to San Juan Harbor, one of the Caribbean's busiest ports.

Eight of the barge's nine cargo tanks were damaged in the initial grounding. During the first days of the spill, some 598,000 gallons of No. 6 fuel oil poured into the ocean. On January 15, 1994, after much of the oil was removed from the fatally crippled barge, the USCG, with help from the U.S. Navy Supervisor of Salvage (SUPSALV), floated the vessel out to sea and sunk it 6,600 feet under water at the U.S. Navy explosives dumping area 20 miles NNE of San Juan. Towing and sinking the *Berman* caused a subsequent major release of almost 200,000 more gallons of oil – one more step in a continuous event of oil leaking from the barge – bringing the total oil spilled to an egregious 798,000 gallons.[1]

These are the basic facts of the *Berman* disaster uncovered during an intensive three-day hearing conducted by the USCG.

The story begins much earlier.

FAULTY THINKING

How could a barge strike a well-charted reef? Was the tug towing the barge manned by competent, licensed seamen? How could the tug *lose* its barge? Why didn't someone notice the barge had become disconnected? Who was in charge? In short, how could the *Berman* spill have happened?

The answers to questions raised by *Berman* indicate serious flaws in the way the oil shipping industry conducts its business. The owners, operators and charterers of these immense vessels and the companies charged with ensuring the safe transport of oil across the ocean contribute to the problem. The *Berman* spill was a clear case of "whatever can go wrong, will go wrong." A series of mistakes and poor judgments led to this 1994 spill. At almost every step

Waves of spilled oil from the *Morris J. Berman* wash up at Punta Escambrón. (Reprinted by permission of U.S. Coast Guard MSO San Juan.)

along the way, people in charge made decisions that put the vessel and, by extension, the marine environment at increasing and unnecessary risk.

The real cause of the disaster was incorrect *bottom line* thinking. Choices were made to save a little money by skimping on maintenance and upkeep of essential equipment, shopping around for the best buy in towing wire, skipping a few safety details, getting by on one more use out of a badly deteriorated towline, and running a tugboat on a small crew with a tug captain who had little or no sleep in the crucial hours before critical decisions were made.

Berman provides a glaring example of the worst factors in the world shipping industry, of which the International Maritime Organization (IMO) has warned: an older, single-hull vessel, skeletal crews, overworked captains, concern for profit margins far outweighing safety and the environment, and gaps in training crews in maintenance and shipboard procedures such as emergency response. All these elements worked together to cause the *Berman* spill.

We have all been guilty at some time in our lives of faulty thinking in a crisis, of making a wrong decision first and then complicating matters by proceeding along that path, rather than retracing our steps or starting over. Even the most logical, well-meaning people may stumble, take increasingly wrong turns, and end up hopelessly lost. The players in *Berman* were deep into faulty thinking as early as August 1993. Some five months before the final events leading to the ultimate demise of the barge, the principals in charge made one poor decision after another. Any one decision made differently could have avoided the spill.

UNDER THE CARE

An integral link between the owners, operators and charterers of the barge and tugboat lies at the heart of this disaster. On the surface, it appears that different entities owned and operated the two vessels involved in the spill. The *Morris J. Berman* barge is owned by New England Marine Services, Inc. (New England Marine) with its principal offices in Bayonne, New Jersey. Metlife Capital Corp. owned the *Emily S* tugboat, but the boat was under charter to the Bunker Group of Puerto Rico, Inc. (Bunker Group) with its offices in New York. In fact, both vessels were operated by the Berman/Frank families of New York, which is where the connection to Puerto Rico and this company enterprise begins.

The *Morris J. Berman* worked the waters of New York Harbor as a sludge barge until the Berman/Frank families and their businesses, Berman Enterprises, Inc., Standard Marine Services, New England Marine and other shipping and marine-related companies, were prohibited from operating their many vessels in the state of New York.

This extraordinary action was credited to the family management style, characterized by an extreme lack of care for their numerous vessels, including the *Morris J. Berman* and *BGI Trader*. The Berman/Frank conglomerate showed a clear contempt for regulations designed to protect the public at large and the fragile environment of the New York waters. Federal and state agencies, including the USCG, assisted the Commissioner of New York in investigating this family and their chroni-

5

cally poor shipping practices. The investigation resulted in a lengthy hearing and a court order enjoining the company and its subsidiaries from conducting business in New York. The New York Commissioner summarized the horrors of their New York and New Jersey operations:

> *"Their barge-hauling activities showed a pervasive and unabated pattern of illegal conduct, which extends to criminal mishandling of waste materials, wholesale operation of uncertified barges, disregard of Coast Guard enforcement orders, including refusal to pay penalties levied therein, failure to license barges with DEC* (Dept. Of Environmental Conservation) *or report oil spills at critical junctures, and knowing misrepresentations and concealment of material facts in licensing proceedings before DEC."* [2]

The state of New York found that vessels owned or operated by this family presented a clear and immediate danger to the health and welfare of the people of the state, or that the continued operations were likely to result in irreversible damage to the state's natural resources.[3]

The family's poor regard for operations resulted in two petroleum explosions, death, serious injury, and 10 oil spills, including one of 50,000 gallons. The environmental disaster that provided the finishing touch involved dumping raw sewage into navigable waters in New Jersey's Newark Bay and in Brooklyn, New York. Evelyn Berman Frank, the head of the family, known in the industry and the marine community as "the Dragon Lady," received a jail sentence which was commuted to several years' probation and community service. Recently the 79-year-old violated the terms of her probation, and agreed to step down as head of the family, but failed to do so. She was sentenced to serve three years in jail, and her corporation, General Marine Transport Corp., was fined $1 million.

Thomas Sommers, a Berman/Frank marine consultant, confirmed the connection between the Berman/Frank members and their various companies at the New York licensing hearing. He described Berman Enterprises as a "family company" begun prior to World War II in Staten Island. He described how their barges, tank cleaning facility and numerous companies under different names were all "in the care of the Berman Industries."[4]

After being expelled from New York in 1991, this shipping family enterprise moved the *Berman* to Puerto Rico. Pedro Rivera, the Puerto Rican manager and representative of Bunker Group, told USCG investigating officer Capt. Larry Doyle that his company had one primary customer, Caribbean Petroleum Corp. This oil company needed No. 6 fuel oil moved from San Juan to other ports in the Caribbean. The Berman/ Frank's Bunker Group took on the job with three of their vessels: the *Morris J. Berman* barge, the *BGI Trader* barge, and the tugboat *Emily S.*

According to Rivera, *Emily S* originally operated in New York and hauled the *BGI Trader* barge to Puerto Rico for this customer. The New York crew stayed on with their tugboat.[5]

Basically, the Berman/Franks owned and operated both the tug and barge involved in the *Berman* spill. Their operations were generally characterized by bad management, inadequate training of personnel, and an improper or total lack of regard for maintenance of the barges and tugs in question.

The USCG local and district level offices (New York, Miami and California officials and offices were involved in the licensing hearings) were intimately familiar with the egregious behavior of this shipping family, dating back to the New York licensing investigation. In addition to the notice of unsafe practices, the USCG had before it a complete inspection record of the *Morris J. Berman* each time new certificates for operation were issued. Tank vessels, including barges towing flammable or combustible materials such as oil, must have a current Certificate of Inspection (COI) before they can sail. Inspections occur at various stages before COIs are issued. The USCG:

- Initially reviews vessel plans against construction standards.
- Conducts two-year inspections.
- Conducts a mid-period exam to ensure that equipment meets specific safety standards.
- Conducts 2.5-year checks for salt water service (or five years for freshwater service).
- Conducts drydock exams.
- Surveys for any defects that affect safety whenever a ship is involved in an accident.

These examinations are mainly static inspections conducted when the vessel is in port, rather than underway. On barges like *Berman,* many of the areas to be inspected can only be checked when the barge is in a shipyard.[6]

The most current COI for the *Morris J. Berman* before the January 1994 oil spill was issued August 2, 1993. At that time, the USCG's inspection report on the *Berman* read like a who's who of bad business. The same poor standard practices that led to the *Berman* spill characterized the record of review.

The USCG inspected the vessel repeatedly and noted engineering, safety and operational deficiencies. The company performed only partial corrections, and new certificates were issued with the uncorrected defects noted. This pattern repeated itself from January 1991 through December

17, 1993, the last USCG inspection, a scant few weeks before the *Berman* spill occurred.[7]

In spite of the family's record of bad business, the USCG found the *Berman* seaworthy, granted COIs to the *Morris J. Berman* and *BGI Trader*, and allowed these vessels to operate in Puerto Rico.

A finding of seaworthiness focuses on the vessel's structure and equipment. If a ship meets the high standards for structural and functional safety set by the USCG, then a COI is issued. Prior to the *Berman* spill, the USCG viewed its recertification role as limited to determining the seaworthiness of vessels, which did not include evaluating the integrity, business practices and character of the vessel owners. The "human factor" did not enter into the equation. Regardless of how bad the Berman/Frank's business reputation was, if their vessels were found "seaworthy," the COI was issued.[8]

With certificates in hand, the Berman/Frank family seized the opportunity, took their businesses and vessels out of town and continued their sloppy and dangerous activities in Puerto Rico, to the detriment of the people and natural resources of that island, and to U.S. taxpayers.

BOTTOM LINE THINKING

The Berman/Frank family's trademark tight-fisted operation in New York carried over to their Puerto Rico tug/barge business. Profit margin, not safety, pervaded the thinking of those in charge. At each juncture where careful attention and forethought might have made a difference, concern about dollars won out and, in the end, the public lost.

The first major poor decision in the oil spill sequence of events involved a choice of money over safety. On three separate occasions, the agent or captain in charge could have replaced the worn and corroded towline that eventually split and caused the barge to separate from the tug.

According to an FBI metallurgist, the towline between the *Morris J. Berman* and the *Emily S* was a wire rope 1.4 inches in diameter made of galvanized steel. The towline flowed from a drum on the rear of the tug, passed over two vertical rollers, and crossed through an anti-chafing plate to the towed barge. When the towline broke for the first time during the *Berman's* last fateful voyage, an emergency repair was made and the line from the tug was connected to a 30-ton shackle on the barge's emergency wire rope line.

The sea environment is tough on a carbon steel wire rope. The line is subject to constant exposure to rain, condensation and salt water spray. The wire spends considerable time immersed in salt water, resulting in wear, tear and corrosion, even when a company practices good maintenance. Where upkeep is poor, the tow wire's deterioration occurs even faster. According to the FBI expert in the *Berman* investigation, a towline is much more complicated than its outward appearance might indicate:

> *"Wire rope is a complex 'machine,' composed of a large number of precise, moving parts which are designed and manufactured to bear a very definite relation to each other. In fact, wire rope has more moving parts (approximately 252 in this case) than most mechanisms falling within the broad category of 'machines.'"*[9]

Proper maintenance includes lubrication of the rope to reduce abrasion and prevent corrosion from the elements. Lack of such care leads to brittle, corroded and unstable wire rope, subject to failure in the "wrong" circumstances. The *Emily S* engineer admitted that he *never* lubricated the towline once during its entire life since its installation in November 1991. The only service he performed was rinsing this steel wire with fresh water after each use.[10]

As early as August 26, 1993, when the towline parted the first time, 80 feet from its end, the Bunker Group could have purchased a new line and avoided the *Berman* spill. A new towline costs about $9,000. The line was repaired, not replaced, by Astro Industrial Supply, Inc. in San Juan, who threaded the wire through a thimble device to form a bridle, then molded the wire together with a molten lead-like substance and installed a socket fitting.

On December 8, 1993, the same towline split again on a trip to Guayanilla, on Puerto Rico's southern coast. The line could have been replaced at that time. In fact, Kehoe, the tug captain in charge on that date, stated emphatically to Pedro Rivera, the company's agent, that "the tow wire must be replaced."[11]

Rivera admitted that Capt. Kehoe told him on several occasions that a new wire should be purchased, and that Kehoe stated, "When we come back from Guayanilla, I want this cable installed."[12] The tug returned from Guayanilla December 26, 1993. Unfortunately, Kehoe was on vaca tion when McMichael took charge of the tug *Emily S* in the days leading up to the oil spill.

Pedro Rivera chose to ignore the advice of his tug captain and did not replace the defective towline because of "cost considerations." Rivera was price-shopping, trying to get the best deal.

> *"What happened is, I am in the process of purchasing a new tugboat in Louisiana and we were considering bringing the cable from the states. We looked at the alternative of pricing, buying it locally versus buying it in the states, and it was a consideration. You know, when I buy everything I go through the process of checking on what are my best alternatives."*[13]

Finally, Rivera cut a check for a new tow wire December 22, 1993, but *still* the company failed to install the new line. Again, cost was cited as the reason. Bunker Group wanted to squeeze one more trip out of the already frayed wire. Rivera bought the new tow wire, and it was ready to be installed December 22. The towline itself was in the hands of the local supplier and installer, Astro Industrial Supply.[14] A supply representative told Rivera that replacing the new towline would take about two hours.[15] The work could have been performed at any time between December 22, 1993 and January 5, 1994, but Bunker Group refused to stop their operations long enough to make the replacement. Instead, they kept running the barge and tug from Guayanilla in southern Puerto Rico to San Juan between the date the towline was purchased and the tug's return to San Juan December 26. Apparently Capt. Kehoe started his three-week vacation shortly before December 26. Pedro Rivera also took a few days off for Christmas, and returned December 28.

Even in their absence, the line could have been installed. Two engineers remained on board the tug. Rivera claimed the engineers wanted rollers on the towing winch replaced before a new line was installed. However, these rollers were repaired December 28. The tug sat idle between December 28 and January 1, 1994, and still no installation took place when McMichael finally returned to the boat. For some unknown reason, the company failed to change out the defective towline, and the new line remained in storage. The tug worked until the morning of January 2, and then sat idle again until mid-morning January 3.

The tug *Emily S* returned to San Juan in the early morning hours of January 6, 1994, when Pedro Rivera and McMichael discussed the sailing orders. The tug was scheduled to leave that afternoon for a trip to Antigua. When they discussed the issue of the new towline, Rivera informed McMichael that January 6 was "Three Kings Day," a national holiday in Puerto Rico when no one works. (In Puerto Rico, even the local grocery store, which is *always* open, closes for Three Kings Day.) Astro Industrial Supply told Rivera they would be happy to replace the tow wire January 7. Rivera never considered asking the other local supplier if they could perform the job January 6.[16]

Scheduling and money concerns ruled the day. The contract between Caribbean Petroleum Corp. and Bunker Group contained a penalty clause for product delivered in Antigua below a certain temperature. If No. 6 fuel oil cools too much, it will not flow through the underwater pipeline connecting the ship to the shore-based facility. Rivera believed if he waited for the towline installation, the barge would not get underway until late in the day. This would equate to an 18 to 24-hour delay from the time the oil was scheduled for delivery to the *Berman* January 6 until the barge actually left port January 7. Such delay would cause a drop in the fuel oil temperature.

Rivera made his decision based on assumptions regarding the penalty clause. According to records from the Caribbean Petroleum Corp., had he bothered to check with the company, they could have accommodated an alternative loading schedule by simply delaying the loading. The fuel would not have sat cooling in the hold while the towline was replaced, and the merchandise could have been safely delivered within the next few days. Bottom line thinking eliminated any consideration of such alternatives. Bunker Group wanted to avoid a possible penalty at all costs and chose to sail January 6 rather than wait one day to install the new line.[17]

McMichael and Rivera decided to gamble. They took a chance that the frayed and deteriorated wire rope would last one more trip. In making his decision, the tug captain was influenced by the fact that the barge was carrying only half its normal load. Capt. McMichael elected to install the new line after their return from Antigua. McMichael related this exchange with Rivera: "So he asked me, 'Would it be all right to make that trip? Is it real essential that you need the wire?' And I said, 'No, that we could make the trip, come back and change the wire. No problem.'"[18]

"An ounce of prevention is worth a pound of cure." In this case, a $9,000 towline might very well have prevented an $87 million oil spill cleanup.

WHEN THINGS GO WRONG, THEY *REALLY* GO WRONG

The plague of errors continued. When the towline failed the first time, shortly after the tug left San Juan Harbor, about 12:30 a.m. the day of the spill, McMichael made a series of decisions that led to disaster. The tug recovered the detached barge, and the tug captain decided to make an emergency repair to the towing wire without calling for USCG or other help. This violates safe standard procedure and federal law requiring any vessel carrying oil or a hazardous substance to contact the USCG whenever there may be even a *threat* of, or potential for, a spill incident. McMichael elected not to return to the safe harbor of San Juan, but to proceed on to Antigua. When the wire parted that first time, the tug and barge were only two miles offshore. A second tug could have been summoned and on the scene within an hour.[19]

McMichael also chose not to ask for help from George Emanuel, a "tug Master" (Operator of an Uninspected Towing Vessel), even though he was in direct radio contact with Emanuel, a longtime local familiar with the Puerto Rico coast. Emanuel was on shore at the port and spoke with McMichael earlier about the first towline break. He and the tug captain were friendly with each other. Emanuel knew the tug *Emily S* and had made a trip to Guayanilla with McMichael prior to December 1993. He noticed then that "the cable showed wear… the wires were bad," and told McMichael, "You need to change that."[20]

Emanuel offered his assistance at the time of the first towline break, speaking with McMichael by radio from the marina. McMichael needed another master's repair knowledge. In spite of his 20 years of experience as a licensed master, McMichael had repaired a towing wire only once before this incident. Yet he ignored his more experienced friend's offer of help on how to make the repair, and his advice on whether or not to proceed.[21]

Pedro Rivera was in the marina on the dock at the same time, and overheard George Emanuel's conversation with the tug captain. When Rivera solicited Emanuel's opinion about whether or not to return both vessels to port, Emanuel deferred to McMichael's decision to proceed. Emanuel did offer Rivera some much-needed advice, which neither Rivera nor McMichael followed: "The tow wire is already set tight, it's tender. It needs to be babied. You need to back off and to slow down to get that load to 'stockhold' in safety."[22]

CASE OF THE MISSING THIMBLE

The manner in which McMichael made the emergency repair compounded an already bad situation. Although the tug captain believed he had "ample time" to make the repair the first time, apparently he was in too great a hurry to do the job right.[23] Crew member Brian Forry and the captain fastened the barge's emergency towing line to a repair made at the end of the parted towing wire. This ineffective repair turned out to be the straw that broke the camel's back and subsequently caused the spill.

McMichael and Forry incorrectly fashioned a *soft eye* by bending the tow wire back on itself in a loop and fastening it with five clamps to the shackle, which then connected to the wire rope on the barge. The customary and only safe practice for such a repair is to insert a simple device known as a *thimble* into the soft eye. The thimble distributes the load evenly, reduces stress on the shackle, and is not kinked so hard that it breaks. In contrast, when the soft wire is bent back on itself in a hard radius turn, without the thimble, the "repair" can break the individual metal strands in the wire. Shortly after McMichael made his soft eye repair, the towline parted for the last time.[24]

The necessary thimble was on board. When Forry rushed to the storage locker for repair equipment, he simply grabbed the clamps and failed to pick up the thimble, which was stored in the same locker. The tug captain knew the thimble was on board, knew how useful and important it was, and in his haste to effect the repair simply "just didn't really think about it in the hurry and scurry in getting the wire together." The tug captain called this catastrophic mistake an "oversight."[25]

Later, the metallurgist determined that the wire would not have parted a second time if the thimble had not been missing: "The catastrophic failure may not have occurred when it did had the wire rope eye been

REPRESENTATIONAL ILLUSTRATION OF TOWLINE ASSEMBLY – *BERMAN* EMERGENCY REPAIR

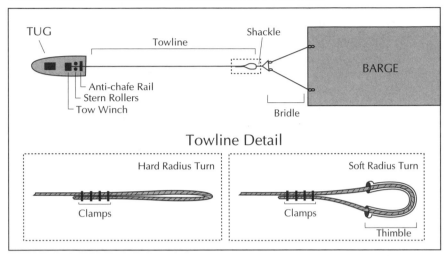

Towline Detail

A towline clamped without a thimble creates a hard radius turn which can cause individual wire strands to bend and break.

configured with a thimble."[26] Capt. Robert Ross, USCG, the Federal On Scene Coordinator who commanded the spill response, later pointed out that the wire was "shot," with or without the thimble. If the thimble had been used, McMichael might have been lucky. Even with the thimble, the repair might have failed an hour, two hours, 24 hours later, or not at all. In the oil shipping business, we should depend on sound management rather than luck.[27]

Brian Forry urged the tug captain to head back to port after the repair. McMichael ignored the seamen's warning and continued on his way to Antigua.[28]

"I LOST MY BARGE"

There was still a chance to avoid the accident, but for two more decisions made by the tug captain. George Emanuel assumed that any competent tug master would "baby" the tug along, slowing down the speed to avoid undue friction or whipping on the fragile towline. McMichael failed to take that precaution. Shortly after the tug resumed its course for Antigua, McMichael took the next two fatal steps. He set the autopilot for Antigua and gradually brought the throttle up to full power, about 4.5 to 5 knots with a half load.[29]

Then, incredibly, the tug captain left the bridge and let all other crew members go back to sleep except for his relief pilot, Victor Martinez. During the course of the emergency repair, Martinez woke from a sound sleep and took the wheel. He saw all that happened and was aware of the lack of thimble, the shaky condition of the tow wire, and all the other circumstances surrounding the emergency. While Martinez later told the hearing officer that he would not have proceeded at full speed if he had been in charge, he did not question the tug captain's orders. Although Martinez assumed control and was the junior officer in charge when he was at the helm, he simply deferred to McMichael and proceeded at full speed. He did so even though he considered such high speed to "not be safe."[30]

McMichael's parting words of warning to Martinez were, "Just keep an eye on the barge."[31] The tug captain, deck hand and engineer watched the repaired towline for no more than 15 or 20 minutes and promptly fell asleep.

Possibly the tug captain's lack of good judgment could be blamed on the fact that he had been awake for 17 hours of the 24 hours preceding the grounding. He was overworked, tired, poorly trained in emergency repairs, concerned about avoiding a penalty clause, and unequipped to handle a major emergency like the one which confronted him at 12:30 a.m. January 7, 1994.

Apparently Martinez was not in the best condition at the time either. He failed to heed the tug captain's order. He "woke up" about 3 a.m. when Rivera called from shore to ask for a progress report and to see how the repair was holding up. Martinez glanced behind and saw one running light on the barge, the red light. He failed to appreciate the fact that he could not see the barge's other green light at the time. Unknown to Martinez, the towing wire split again at 2:30 a.m. and the barge was already drifting toward shore. It was not until 4 a.m. that Martinez woke up Brian Forry and asked him to help find the barge that was lost for the *second* time that night. The normal distance between a tug and barge depends on sea conditions, but is usually about 300-400 feet – about the length of the *Berman* itself. At the time of the grounding, the *Emily S* was *11 miles* away from the *Berman*.[32]

Martinez did not notice this problem because he failed to stand a proper watch. In simple terms, Martinez fell asleep on duty. Perhaps Martinez's actions (or inactions) can be explained additionally by the fact that he had consumed cocaine at some time prior to the incident. The crew was tested for drug use when they finally returned to San Juan Harbor about 6 a.m. January 7. Forry tested positive for marijuana and Martinez tested positive for cocaine in the amount of 290 nanograms, not an insubstantial quantity.

Why didn't these companies, engaged in the hazardous business of transporting oil, determine if their employees abused illegal substances? We can only speculate that cost considerations entered the picture. Neither Bunker Group nor New England Marine tested their employees for drug use as required by federal regulations.[33]

BOOMING SEA

Two tankermen were on the barge when she struck the reef at Punta Escambrón. Although they were not supposed to be on a barge certified as "unmanned," Philip Taubler and Scott Bruce, able-bodied seamen and tankermen, elected to travel to Antigua by barge so they could conduct some needed repairs.

Taubler awoke when he heard waves breaking on the barge and sea water slamming against its sides. He tried unsuccessfully to contact someone on the tug, but his radio communication failed to elicit any response, probably because Martinez was asleep. Next, he called the USCG. Having observed oil gushing from the barge, he says that he asked for immediate booming to protect the beaches. He shouted for his mate and waited for help.

Scott Bruce woke when he heard Taubler yelling that they were in trouble. He ran to the front of the barge and dropped the anchor immediately. He described the nightmarish scene:

> "The anchor dragged a little bit as the seas were crashing all over us, and now I'm saying it was a two to four-foot sea out in the ocean, but as you're coming closer to the breakers, it builds a lot bigger … The smell of the oil was so bad that I – you know, it was unbelievable and it was burning my eyes – that I knew we had a serious problem."[34]

Both tankermen agreed that if they had had enough warning and had known about the second break, they could have rigged another line between the barge and the tug.[35] The final opportunity was lost. Martinez failed to give a timely warning.

Error upon error, mistake after mistake and bottom line thinking resulted in catastrophe.

WHO'S TO BLAME?

The obvious culprits are the individuals who made one bad decision after another: Pedro Rivera, as agent for Bunker Group and New England Marine; Roy McMichael, as the man in charge of the tug; and Victor Martinez, as the last individual on watch during the final hours before the spill. Equally easy to blame are the Berman/Frank companies, whose greed manifested itself in poor training, minimal equipment maintenance, cost-cutting wherever possible, down-sizing crews so that captains and deck hands were overworked and on the edge of fatigue, and not maintaining drug-screening programs.

Some might argue that the USCG bears responsibility for certifying the Berman/Frank vessel *Morris J. Berman* as seaworthy in the face of the family's disastrous environmental and poor safety record. At the time of the incident, the focus of the USCG was on seaworthiness, and not on the value systems behind the operators of the tug and barge. Still, there were deficiencies on the record that could have been cause for denying the renewal of the last Certificate of Inspection.

There is an even deeper level of inquiry: the government, through its regulatory agencies, has failed the U.S. and commonwealth citizens. The operators of towing vessels, like Roy McMichael and Victor Martinez, are licensed by the USCG. The licensing procedure is fraught with glaring deficiencies. The tests given for a masters or mate's license do not question the operator about towing equipment maintenance, adequate size of towing gear, or emergency repairs and procedures. The regulatory experience needed to qualify does not require individuals to have specialized training or schooling in towing operations and equipment. A minimum of one year experience on the deck of a tug is all that is required for one to hold the type of license Roy McMichael or Victor Martinez possessed. As was evident from the *Berman* spill, these individuals were poorly prepared to deal with the towing line emergency they faced in the early morning hours of January 7.[36]

At the time of *Berman* the USCG did not exercise any oversight or control over towing vessels. The only towing vessels inspected were sea-going vessels over 300 gross tons; all others are exempt from inspection. Less than 10 U.S. towing vessels fall in this category. This meant that no agency was in charge of determining the safety and adequacy of vessel gear, such as the towline, shackle, thimble setup and other towing equipment at fault in the *Berman* spill. This task was left to the companies operating the tugs and other towing vessels. An irresponsible business could simply ignore such duties, although at some peril of criminal and civil sanctions. Almost two years after *Berman* the USCG finally passed a regulation asserting controls over tank barges and the towing vessels hauling them.[37]

Capt. Robert Ross, USCG, supervised the *Berman* response and argues that all the regulations in the world would not have prevented this disaster. "How do you regulate stupidity?" he asked.[38]

He has a point. There will always be human error. But a minimum inspection system could have disclosed the need for a new towing wire and prohibited the Emily S from sailing without replacing this critical part of the towing system. Capt. Ross argues that the inspection intervals would have to have been just right to catch the problem and that the only guarantee is with the managing agent of the tug who knew the towline was frayed and should have replaced it. Proper training of the captain and crew should have better prepared them for repairs and safe procedures.

There are other less obvious perpetrators, such as the legislators and special interest groups who killed badly needed legislation intended to correct these deficiencies. *Berman* was the final incident which led Rep.

Gerry Studds to propose the Towing Vessel Navigational Safety Act in 1994. This act recommended rigorous licensing for towing vessels, entry level training for deckhands, third party inspections and much more. The USCG's study, released in 1994, supported the bill, which died in the U.S. Senate. The future of such legislation is unclear, given Studds' decision to leave office and the disbanding of his Merchant Marine and Fisheries Committee. In May 1996, Sen. John Chafee proposed legislation that addresses some, but not all, of the towing vessel problems. Again, the fate of this bill is unclear.[39]

The *Berman* tragedy is ripe for a repeat performance. Between December 1993 and year-end 1995, the USCG in San Juan, P.R. faced three other tank vessel groundings, one a barge and the other two ships. The barge *Superior*, like *Berman*, involved the failure of towing gear known to be defective but not replaced. Unlike *Berman*, the *Superior* was double-hulled and lost only *10 gallons* of the 600,000 aboard.[40]

We must look even deeper for the ultimate wrongdoer. It is painful to admit that all of us who consume oil products and byproducts bear our share of responsibility. Our insatiable fuel demands make the business of shipping oil profitable. Until we shift our consumer needs, demand less energy and seriously consider alternative sources of energy, bottom line thinking will reign supreme. Unless we demand strict and comprehensive regulation, and insist that laws now on the books stay in effect, oil will travel unsafely. Otherwise, polluters may continue to cut corners, try to avoid paying damages, or simply pay the cost and get on with their scheduled business.

Morris J. Berman aground and discharging oil. (Reprinted by permission of U.S. Coast Guard MSO, San Juan.)

THE WAR ON SPILLED OIL

IT'S WAR!

The *Berman* oil spill response took place between January 7 and April 30, 1994, the official last day of the cleanup. Those involved repeatedly described the response effort as a "war."

Oil spill responders use a language of classic military textbook jargon: *Hazardous Duty, Salvage, Decon, Pseudo Military, Incident Command System, Unified Command, Sam Mission, Battle Staff, Situation Room, Small War Room, Scat Teams, Turf Battle, Ramp Up and Ramp Down, Spill Specific Cascade Schedule, Shoot First, Ask Questions Later, Sneak Attack, Safety Envelope, and Active Triage.*

The most effective way to fight an oil spill is to wage war against it. "Battle grounds" include the ocean surface, middle column of the water, sandy bottom, rocky shoreline, beaches, mangrove swamps, seagrass beds, and other fringe areas bordering the coastline. The front line victims are the marine plants and wildlife surrounded by, touched by, or lying directly in the path of the oil. The invader is an all-pervasive substance that changes its character from dissolved oil droplets to thick, black tarmat sheets and every consistency in between, given time and the elements.

How do you combat an enemy like crude oil? It may float in tarballs on the ocean surface, dissolve in the water column becoming a brown, mousse-like substance, or descend to the bottom of the sea killing certain animals on contact, irreparably injuring others, and disrupting entire habitats. When you think you have defeated this enemy, petroleum may resurface miles away from the spill, or recoat a recently cleaned beach. The foot soldiers are the troops on the beaches, in helicopters or small airplanes hovering over the water, on ships on the surface, and diving in the sea. The officers in charge wage their campaign with a clear objective: clean up the spill, provide the greatest protection possible for endangered flora and fauna, and cause the least harm to the people performing the job.

An oil spill cleanup is different from an actual war in at least one obvious respect: no one is shooting back – at least not with traditional weaponry. "An oil spill is the closest you'll come to a war without killing someone," noted the president of one of the major response corporations involved in *Berman*.[1] Oil is a lethal killer, and the battlefield of slaughter is determined by the destruction oil causes to all plants and animals anywhere near its path. Quick death by suffocation and toxic ingestion is the likely result of initial contact. Ecosystem victims spared immediate death often face slow poisoning from chronic, low-level concentrations as oil rereleases as long as 20 years after the incident.

In war, commanders plan their attack strategy in advance of the actual conflict. A good battle chief often carries out the tactics he has learned from careful study and experience, develops major defenses, attacks, counterattacks, and deals with the logistics of battle by following certain defined principles.

Oil spill response rarely matches the hypothetical working models developed in advance. Prior to a spill, many individuals in the community, working with specialized agencies, devote considerable time and attention to developing an Area Contingency Plan (ACP). The ACP tries to predict what may happen when a spill takes place, sets priorities for protection, and then parcels out areas of responsibility among the players. But the spill itself is a surprise attack, a Pearl Harbor "Day of Infamy" against the environment. The spill takes everyone off guard, and few detailed plans are ever executed as written. The only certainty is to *expect the unexpected.*

Among the unpredictable factors that seriously affect any ACP are:

- Where the spill occurs.
- The type of oil released into the ocean.
- How quickly and over what size area the discharge has dispersed.
- Techniques required to adequately clean up the discharged oil.
- The condition of the culprit vessel.
- The owners of the ship.
- The authority charged with responding.
- Weather conditions affecting the response.

In spite of this uncertainty, advance response planning for the "next" spill is valid and valuable. The most important product of an ACP is not the scheme itself, but the act of designers working together to assemble a response team. In this process, team members:

- Meet other response members.
- Learn who to contact.
- Establish how to best coordinate their efforts.
- Identify ecologically sensitive areas.
- Prioritize cleanup sites.
- Formulate an overall response strategy.

The detailed plan is less important because the actual situation is typically nothing like the hypothetical model. During an actual response, written plan details may simply fly out the window. The response then proceeds as planned on the ground at the time of the actual event, using the tools and people at hand under the circumstances dealt by the spill.[2] In *Berman*, major stakeholders ultimately were "very pleased with the speed and thoroughness of the cleanup," possibly because of the advanced planning.[3]

SHOOT FIRST, QUESTION LATER

The similarities between war and oil spills are startling. "There is no easy oil spill," says CDR Edwin Stanton, USCG, Capt. Ross' operations director.[4] (Capt. Ross was a commander at the time of the spill and was promoted to captain shortly after the cleanup; CDR Stanton was a LCDR at the time of the spill.) The method of tackling the task is the same: Get there first with the most. Time is critical. The rapid breakdown of oil in ocean water and its increasingly toxic effect on flora and fauna calls for immediate response.

In the *Berman* spill this meant shoot first and ask questions later. *Berman* was a highly publicized incident, and the public demanded quick response. Capt. Robert Ross, USCG, the officer in charge (Federal On-Scene Coordinator, or FOSC), took the most appropriate action commensurate with the risk, but in rapid fashion. Normally the *Responsible Party* (RP) would be in the forefront of the cleanup. Due to prior experience with New England Marine Service, the *Berman* RP, the FOSC knew he had to act quickly and not wait for them to take the lead. This resulted in a very aggressive government response.[5]

Oil spills and actual battlefields are both highly charged emotional experiences. Those directly involved in *Berman* described the oil adrift in the ocean, smashing against rocks, layering beaches and the floor of the sea, as "horrendous," "devastating," "toxic," "a highway of asphalt," and "sickening." The effect on responders and volunteers in, around or near the spilled oil differed, but many experienced fear, anger, despair, anxiety. You don't reason with oil, you simply deal with it. Decisions are made based on shifting weather conditions, ocean currents and other natural forces beyond man's control. Tension and fatigue are constant.

Individuals in charge find themselves in highly volatile situations without clearly defined rules, from which they must resolve a multitude of difficult problems for which there are no easy answers. For the front line fighters, the FOSC and his immediate staff, the sphere of operations is a series of grim skirmishes and pitched battles for control over an elusive enemy. One participant described command headquarters as "organized chaos."[6]

Finally, as in military action, oil spill responders commonly use a bewildering array of acronyms as shorthand for the many organizations, entities and regulations involved in the industry's response and prevention. In this text, acronyms are spelled out in first reference and repeated where helpful. A complete acronym reference is provided in Appendix C.

IVORY TOWER

Battles are conducted from the top down. At least in theory this same strategy applies to oil spills. The U.S. operates from a three-tiered structure. The first tier is the Incident Command System, a model based on military theory descended from the World War I-era Prussian staff system. National Park Service firefighters refined this older system into an *Incident Command System* (ICS) to battle large forest fires. ICS is the current U.S. model for fighting oil discharges. ICS divides the work into manageable, yet related segments that provide the *how* of the response:

- *Command* is the authority that prepares a plan of action.
- *Operations* stages and implements the plan, directing the how of the response, for example, surface, air, shoreline.
- *Planning* gathers and analyzes information to evaluate and revise the plan when necessary.
- *Logistics* provides backup support, food, equipment, and more.
- *Finance* pays the cost.

The second tier is the *Unified Command Structure* (UCS) which typically involves federal, state and local agencies and the Responsible Party (RP). The UCS identifies the parties involved and provides the details on how they work together, with the FOSC in charge. This is the *who* of the response.

The third tier involves coordination of the prior two with the details of *what, where, when, by whom* and *with what* developed under the Unified Command Structure, (the *who*) in conjunction with the *how* of the Incident Command System and based on the specifics of the incident.

In theory, ICS equates to top-down management. Those primarily in charge communicate through clear lines of authority to supervisors, who in turn direct and manage field responders in each key area of response. Every agency and individual involved has a preassigned place in the hierarchy. Responders complement each other. Those in charge of each area report up the chain to their supervisors. The theory under ICS is that "all horses pull in the same direction."

ACPs contemplate that the RP provides both an Oil Spill Removal Organization (OSRO) and a Spill Management Team (SMT). The OSRO produces the workers and the equipment, while the SMT determines the workers and equipment needed, what tasks are to be accomplished, establishes a work schedule, and other similar on-the-ground decisions. The SMT becomes part of the core of the ICS system with the various agencies working into the overall management plan. (See Appendix A, *ACP UCS Organization* and *Berman Response Management Organization*.)[7]

The *Berman* spill revealed the vast difference between the ivory tower model and practical application. Those at the top in all sectors – federal, Puerto Rico commonwealth and private industry – questioned how well the ivory tower model actually functioned in the *Berman* crisis.

THE MONSTER

Berman provides an excellent case study for how to fight a large discharge of highly viscous material in a high profile or sensitive environmental area. While there were problems unique to this discharge, there are also many similarities to other spills that highlight common issues requiring further evaluation.

On another level, *Berman* emphasizes the harsh lesson that we as consumers have created this oil monster and unleashed it on the world. Humans, animal and plant kingdoms all suffer from the mounting loss of natural resources and habitats. Maybe you will not live to see the oceans become too choked with oil and other human waste to sustain life, but your children or grandchildren could very well face that bleak future. Current scientific evidence suggests that the world's coral reefs are living on borrowed time. Under present conditions, due primarily to man-made pollution activities, two-thirds of all known reefs may collapse within our

Responders battle released oil. (Reprinted by permission of U.S. Coast Guard MSO, San Juan.)

lifetimes, or those of our grandchildren.[8] The *Berman* spill also provides a prime example of the tremendous difficulties involved in trying to put the oil genie back in the bottle once it has escaped – the bottle, in this case, being a ruptured tank vessel.

Berman was the first major spill response effort in the United States since the effective date of the Oil Pollution Act of 1990 (OPA 90), August 18, 1993. The cleanup took place over 114 days. The unique situation and location of the spill required a massive coordination of effort. Fifty contractors and 15 federal and Puerto Rican agencies devoted approximately 1.5 million work hours in the field. (This number excludes the number of sleepless nights many of the command personnel suffered.) At the height of the crisis, nearly 1,000 workers battled the spill.[9]

What was the result of this immense undertaking?

On one hand, the cleanup can be judged a success. Of the 1.5 million gallons on board the *Morris J. Berman* barge when it ran aground, 1.3 million gallons were accounted for, either transferred to another vessel by a procedure called *lightering*, evaporated into the environment, or recovered from the ocean. This leaves some 200,000 gallons of oil awash in the environment, clotting the ocean bottom, trapped inside the sunken *Berman*, caught in the rocky shoreline, or buried beneath sandy beaches.[10]

Major tourist beaches were cleaned and ready for public use again within six weeks, and the worst oiled beaches were restored within three months to the day after the spill. Initial data indicate that natural resource damages and adverse economic impacts were significantly reduced because of the prompt and thorough cleanup. The public felt that those in charge acted responsibly and with appropriate sensitivity to the crisis. Most of the responders involved rose to the occasion and suspended their own bureaucratic needs for the good of the cause.

Yet the *Berman* response was also the "most expensive federal response undertaken." Total cost was driven by two factors, the size of the spill and the opportunities presented to recover so much of the spilled oil. The cleanup cost the USCG (and ultimately the Oil Spill Liability Trust Fund) some $87 million, (although $5 million was later reimbursed by the *Berman* insurer from the total of $10 million the insurer paid out.)[11] Future damage to the environment is still unknown. According to Dr. I.C. White, managing director currently consulting with the International Tanker Owners Pollution Federation, Ltd. (ITOPF), only 10 percent of the original volume of oil spilled is ever recovered at sea. The remainder contaminates beaches, accumulates in salt marshes and mangroves, or ends up in landfill treatment facilities as untreatable waste.[12] The aggressive *Berman* response may have increased the recovery percentage considerably, but the fact remains that much of the oil *spilled* in the environment *stays* in the environment, regardless of how effective the response may be.

As a test case under OPA 90, *Berman* raises serious questions about how much future spills will cost, and how much *you*, the ultimate payor, will be required to pay to clean up the mess. *Berman* exposed glaring deficiencies in how the U.S. and the rest of the world plan for spill responses. Loopholes dilute regulations dealing with certain types of vessels, specifically uninspected towing vessels like tugs and tank barges, active in transporting oil. Tank barges are large storage tanks pulled or propelled by tugboats. According to one study, tank barges were responsible for the highest incidence of oil spills in U.S. waters between 1985 and 1991. Until recent regulation and USCG emphasis on prevention through people becomes effective, these vessels will surely be responsible for more *Berman*-type spills.[13]

The international safety nets controlling oil discharges appear to be failing. To remedy this situation, the international community and U.S. agencies are acting to regulate the shipping industry. People are becoming a more important part of the solution. Shipowners appear willing to take the necessary steps to ensure the safe transport of oil at sea. The corrective process is slow, and meanwhile the seas continue to suffer.

The *Morris J. Berman* spill killed thousands of creatures, from fish and marine mammals to sea birds and invertebrate organisms. Humans suffered significant economic loss for which some, like local fishermen, may never receive just compensation. The coral reef ecosystem it struck was almost obliterated.

Berman can happen anytime, anyplace, and will surely be repeated, given the current state of legislation in the U.S., the attempt to back off from OPA 90's message that *the polluter pays* and wrong *bottom line* thinking that directs some national and international companies in the business of transporting oil.

Future methods for oil spill cleanup may vary. The results will be better or worse. Most people will rise to the occasion. Others may again take advantage of the spill by flushing oil from their vessel bilges during an event because they can do so undetected. The description of the equipment and techniques, manpower (and woman power), ingenuity, inspiration and perspiration, and the considerable cost spent to clean up the *Berman* spill, all reflect similar oil spill casualties that happen almost daily around the world – thankfully, most times in much smaller quantities.

In a larger sense, the 1994 *Berman* event is timeless. The following account may provide a most useful baseline model for battling and even improving oil spill response and prevention in the future. The ultimate hope is that the lessons of this accident will be remembered so we will not be doomed to repeat them.

OFFSHORE RESPONSE AND SINKING THE *BERMAN*

EYES ON THE BOTTOM

As Captain of the Port in San Juan, Puerto Rico, Capt. Robert Ross, USCG, was the pre-designated Federal On-Scene Coordinator (FOSC) under the National Contingency Plan. His first priority was to assess the extent of the damage to the *Morris J. Berman,* determine the ship's condition, and find out how much oil still remained on board the crippled vessel. Even before recovery efforts could begin, the first order of business was to prevent more oil from spilling into the marine environment.

Kevin Peters from Miami Divers was chosen as the primary diver for the vital if unenviable task of assessing the barge's damage. Peters and his brother operate an independent diving business specializing in underwater salvage and cleaning operations. Peters grew up in Minnesota, where he dived under the ice to recover lost cars. He learned welding in the Marine Corps, and later attended the Coastal School of Commercial Diving in San Francisco, worked on oil rigs in the Gulf of Mexico, and finally established his business in Miami. Peters described his arduous experience diving the *Berman:*

BARBARA E. ORNITZ: Kevin, what was your job with *Berman?*

KEVIN PETERS: I was their eyes on the bottom, at first. That means every day I'd dive under the boat and film the underside to see how bad the damage was. That wasn't easy.

Picture this: No. 6 oil is very thick. It's what's left over after diesel and gasoline are removed. This makes for almost zero visibility in the water, where the waves are breaking over the barge and stirring up the oil. While I was videotaping this scene, I had to feel my way along the sides of the barge. It's sort of like "swimming by Braille."

We wear special gear, a Viking drysuit that keeps you completely dry and protected from the high sulfur content of the oil. I drape a full hard helmet around my head and all the way around my neck. This piece of gear holds all my communications equipment. I'm con-

nected to the surface by what I call my "umbilical cord," a line taped to my helmet which runs up top. I can use this if I have to bail out fast, and also when I exit from my dive. I follow this cord up since I can't see well enough to tell top from bottom.

Three lines are tied into my helmet: my air line, a communication wire, and my lifeline. I run the video camera under water and relay the picture to the surface by my communication wire. Since I can't see anything, there's a guy up top who views the picture. I hold the closed-circuit camera in one hand. With my other hand I grope along and try to keep touch with the barge. The surface picture is much better than what I can see in the oily murk. There's a standby diver on the surface to monitor everything in case I hit a problem. Surface air runs through one of my three lines. I also have a "bailout bottle" on my suit in case my hose breaks.

BEO: What were conditions like under the ship?

KP: The bow of the ship was lodged on a huge coral head. I had to get underneath the bow and the stern of the ship to film all the holes. The only problem is that the ship wasn't settled solidly on the reef. The wave action caused it to move. The peak of a wave would come over the barge. As the trough of the wave returned back to sea, it forced water out from underneath the bow. This created a vacuum that would suck me under the ship. The boat was very "lively" all the time I was diving. By "lively" I mean the boat moved a lot.

There was no more than two to three feet between me and the bottom of the barge. That's why I got a little nervous when the vacuum thing happened.

I dived every day I was there until they sank her. I'd stay about one hour, no more. The main question was, "How bad is she?" The answer was: "Bad and getting worse." Later they wanted to know if she was breaking up. It was my feeling the barge would fall apart if they didn't get her off the reef quickly.[1]

The barge was severely damaged. According to CDR Stanton, five of the nine cargo holds were breached and open to the sea, and each of the *Morris J. Berman*'s nine oil holds was in "free communication" with the others. All the internal bulkheads were broken. With each wave that struck the barge, sea water flowed in, then flushed oil out into the ocean. Scarcely enough of the barge's single hull remained to prevent a direct discharge of all the remaining petroleum.[2]

Peters also had his hands full retrieving equipment such as hoses and tools that fell overboard in the rough seas. Perhaps his most harrowing duty was connecting the hose from the *Morris J. Berman* to the *BGI Trader* for the final transfer of fuel from the crippled barge to the lightering vessel.

KP: I helped make the last connection with the only fuel hose remaining. The seas were too heavy for us to push the receiving barge close enough to the Berman to connect the hose between them. I had to swim a line from the barge over to a smaller boat, which then made the connection. Waves were breaking between 10 and 20 feet at that time, coming in sequences of three to four. My job was to swim this messenger line with new hose running through it to the waiting boat.

My only gear was a mask and fins. I jumped into the ocean, holding onto a half-inch line. The crew on the barge were to pay out 600 feet of that line slowly as I made my way toward the small boat. It was really rough going because the waves were big breaking waves. After about 400 feet of rope was let out, my swimming progress was cut in half because of the weight of the line. Every time a wave came by, I was pulled backward. I swam about 100 yards to the small boat, gave them the line, and then made it into the vessel.[3]

LIGHTERING

Using firsthand information supplied by Peters and others on board the barge, the FOSC turned his responders to the job of stopping the flow of oil. This two-step process involved: (1) removing as much petroleum as possible from the barge and transferring it into tanks on other vessels; and (2) *securing the source* – hauling the leaking barge off the reef and sinking it.

Conditions were extremely uncooperative in the first three days of the response, with 15 to 20-knot winds, and four to six-foot swells at sea that grew to five to eight-foot waves as the days progressed, slamming against the stern of the barge. Water poured over the deck of the *Berman*, tossing response crew members around and sweeping equipment overboard. A tumultuous sea of thick, black ooze floated like dark pudding for hundreds of feet around the stricken vessel as the wind and waves rocked it perilously on the reef.[4]

The assembled response crews took on the urgent business of removing as much remaining oil as possible from the helpless barge to prevent more oil from leaking into the ocean. The barge had lodged itself in the worst position pos-

Lightering oil from the *Berman*. (Reprinted by permission of U.S. Coast Guard MSO, San Juan.)

sible: too far from shore for direct access by boat, and too close to the beach and high surf to place booms around the ship or to use more effective oil recovery equipment. Oil was removed from the barge by *lightering*, a process of siphoning off oil by hose into another vessel. A tug kept the receiving barge, the *BGI Trader*, away from *Berman* in deeper water.

This high-risk undertaking was further complicated by the airlift-only access to the barge. Helicopters from the USN, USCG and Puerto Rico National Guard were employed for this hazardous duty, engaging in more than 50 trips to transport gear and people. Recovery operations had to work around extreme weather conditions, and halted altogether January 11 and 12 due to safety threats posed by waves crashing over the stricken barge.

FOSC Ross assembled an experienced team to accomplish this major objective: Paul Hankins, U.S. Navy (USN); LCDR Richard Hooper, U.S. Navy Supervisor of Salvage (SUPSALV); the U.S. Gulf Strike Team; CDR Stanton, USCG; Kevin Peters of Miami Divers; and their various coworkers.

The USN was involved throughout the spill, given its presence at its Roosevelt Roads Naval Air Station in Ceiba, on Puerto Rico's eastern tip, and the Navy's expertise in ship salvage, shipboard damage control, and diving. The SUPSALV appointed two experienced pros, civilian Paul Hankins, pollution response branch, and LCDR Richard Hooper, Supervisor of Salvage and Diving, both from the USN's Naval Sea System Command.

Hankins, a veteran of many oil spills who saw active duty with the Navy before assuming his current civilian position, oversaw removal of oil from the ocean surface, employing high-tech skimmers in recovery operations. He supervised transport of off-island equipment to Puerto Rico. Most necessary machinery had to be flown in from USN bases and private facilities across the U.S. in cargo planes by private contractors, the USN or USCG. Equipment primarily came from Crowley Environmental Services of Puerto Rico, Inc. (Crowley); National Response Corp. (NRC), NY; and Marine Services Response Corp., stateside and U.S. Virgin Islands. On arrival, the Puerto Rico National Guard, civil defense and local police escorted truck convoys to spill sites.[5] (Sample equipment lists from SUPSALV and NRC are contained in Appendix B.)

By early evening January 8, SUPSALV began pumping oil from the ailing *Berman* to a receiving barge. Surface oil recovery started within hours of the initial grounding using VAC trucks in the Hilton Hotel and Radisson-Normandie lagoons. Actual surface oil skimming farther offshore near the barge, which will be described in greater detail, began two days later. At the peak of his operation, Hankins supervised some 216 responders from USN, USCG, Gulf Strike Team (GST) and SUPSALV.

LCDR Hooper is another oil spill veteran, with 14 years of Navy ser-

Responders aboard the *Berman*. (SUPSALV; reprinted by permission.)

vice, the last three working with oil pollution and ship salvage. SUPSALV is the organization charged as the RP in certain cases in U.S. waters and elsewhere, and acts in other cases as the damage control organization, recovering oil after combat operations. For example, in Kuwait, the Navy was called upon to clean up the enormous amount of oil discharged. Wars contribute a high percentage of all oil spilled in the oceans. During the 1991 Gulf War, Iraqi leader Saddam Hussein released an estimated 2 to 2.5 million barrels (84 million gallons or more) of oil into the environment.[6]

Once it became apparent that nothing could put the *Morris J. Berman* back together again, and that the barge was so completely ravaged that she would continue to leak oil until removed or sunk, SUPSALV's goal became clear. "Our job was to eliminate the source," Hooper said, referring to the decision to sink the barge.[7]

Hooper and his ten-man crew, aided by the Gulf Strike Team, lightered oil off the *Berman* through thousands of feet of floating hose into the *BGI Trader* (operated by the RP, Bunker Group) until the oil became too viscous, like peanut butter, to pump through the hose.

The Gulf Strike Team is one of three USCG units that comprise the National Strike Force (NSF), a select, highly-trained force established solely for the purpose of fighting oil spills and hazardous waste discharges. It is the USCG version of other elite U.S. forces such as the Navy Seals and the Green Berets. The NSF is distributed regionally among the Atlantic Strike Team, the Pacific Strike Team and the Gulf Strike Team, from Mobile, AL. The NSF defines itself as "a highly-trained cadre of Coast Guard professionals who maintain and rapidly deploy with specialized equipment in support of federal on-scene coordinators preparing for and responding to oil and chemical incidents in order to prevent adverse impact to the public and reduce environmental damage."[8]

The Marine Safety Office (MSO) in San Juan requested the Gulf Strike Team's help with response in the early morning hours of January 7. Lt. Jg. Jerry Hubbard assembled his emergency team in Mobile and arrived at the spill 12 hours later. At the height of activity, 40 NSF team members assisted the Navy and others to pump oil from the barge, oversee and conduct skimming operations, set booms in protective positions, manage and conduct onshore cleanup from skimmers on the beach, and provide overall monitoring of the response effort.

"God couldn't have done a better job of putting the spill in a worse place," Hubbard said, describing the scene.[9] The Gulf Strike Team and SUPSALV personnel worked 24 hours a day when weather permitted, lightering some 693,000 gallons of oil/water mix from the barge into the *BGI Trader*.

Once filled, the *BGI Trader* delivered its load of liquid oil directly to the Caribbean Petroleum Company or to the Sun Oil Company facility in Puerto Rico for recovery and reprocessing. *Barge 250-1* was used as a receiving and storage facility for skimmed oil, and transferred its cargo to Sun Oil for filtering and refinement.

Skimmer and boom. (Reprinted by permission of U.S. Coast Guard MSO, San Juan.)

Weir skimmer. (NSFCC/Jerry L. Snyder, reprinted by permission.)

Crews in Ponce, on Puerto Rico's southern coast, mucked out about 2,000 barrels of *sludge*, a thick, heavy semi-liquid choked with debris left behind in the holds of the two receiving barges. Pumps described as "meat grinders" removed the sludge through an elaborate screw system to shove the gunk through hoses into vacuum trucks for final disposal at BFI's Ponce landfill.

Experts describe the skimming operation as "simple, but not easy." Skimming operations took place offshore, onshore, shore-based and in the harbor. Certain skimmers were deployed directly into the water, either in lagoons or on beaches next to large caches of oil. The most effective onshore and shore-based skimmers were incline belt and Archimedean screw pump weir skimmers. Responders used booms to capture the oil in one place for efficient skimming. A boom is a flexible, rubberized product that looks like a string of buoys floating on the surface from a distance. Skirt sealed to this line of inflatable chambers hangs 18 inches below the surface and holds the boom down. Oil collects against this material, making it easier to suction up. Some of the skimmers used in *Berman* included:

Belt Skimmer: Two tugboats pull two lengths of boom behind them to form a large, open-end V-shape that traps surface oil inside the boom. The belt skimmer floats at the apex of the V. The three-foot-wide by six-foot-long belt runs into the water at an angle, lapping up oil which sticks to its surface as water pours off. At the top of the belt, the oil is forced through a *squeegee system* that rings the oil out into a sump tank, which is later off-loaded into the hold of a waiting barge.

Weir Skimmer: Essentially a floating pump turned upside down, which operates on oil corralled by a boom and tugboats. The suction end, suspended by giant mushroom-like floats, slurps up the top half-inch of oil floating on the surface. Captured oil is pumped through a hose into a holding tank through a squeegee system like the belt skimmer. The main weir skimmer used was a DESMI.

Primary offshore skimming was accomplished by the USCG VOSS and Marine Spill Response Corp.'s *Caribbean Responder*.

Vessel of Opportunity Skimmer (VOSS): The VOSS is the USCG's skimmer of choice. It is similar to the weir in operation, except the boom runs off a 40-foot outrigger arm from the boat. The other end of the boom attaches to the other side of the boat, forming a U-shape. A weir skimmer sits at the apex of the U and suctions oil through a hose to the ship.

Caribbean Responder: A 32-foot boat launched with the boom from a 208-foot mother ship forms a U-shape or J-shape in which pockets of oil are trapped and collected. The skimmer is a TRANSREC system that resembles a big funnel with its wide end up. The operator can move the funnel collar up or down from the ship, widening or narrowing the lip depending on the oil cover thickness. The operator also controls buoyancy and the flow rate of oil from the skimmer's suction hose to the ship. This state-of-the-art equipment, developed by Norwegians for use in the North Sea, includes on-deck instruments that measure the flow rate and the oil-to-water ratio.[10]

DIGGING OUT OF A HOLE

Paul Hankins of SUPSALV explained why organizing equipment and labor is one of the most difficult tasks of pollution response:

> *"The biggest problem with any spill is logistics, figuring out what you have and then getting the equipment and those controlling it to the spill scene. There wasn't much skimming machinery on the island. Most of it had to be flown in. The U.S. privatizes spill response. Unlike the European and international communities where national forces manage a spill, in the states the RP is responsible for the cleanup. This makes for a lot of different people showing up with a great quantity of equipment. It's up to the logistics guys to get everything arranged in the right place and working effectively.*
>
> *"Logistics doesn't come easy. You can't organize this type of a response without practice. Practice takes money. In the Navy, we participate in four spill response exercises a year. These cost about $200,000 a shot, use only one skimmer, about 1,000 feet of boom and not much else, and no other forces are involved. One exercise in Alabama with two skimmers and a laundry list of other equipment cost about $500,000."*

The logistics of deploying and using skimmers in *Berman* still did not provide an ideal solution. Normally, once oil is in the water, the best responders can hope for is to "dig themselves out of a hole." Often skimming can only achieve a 10 percent to 15 percent recovery of oil from water.[11] To accomplish even this much recovery requires tremendous effort and coordination of many individuals. While this amount may not seem worth the trouble, in *Berman*, the oil/water mix recovered from skimmed and submerged oil recovery operations amounted to 689,000 gallons, of which 90 percent was estimated to be oil. Compared to the norm, *Berman* skimming operations were termed "unusually effective." The benefit to the marine environment can hardly be calculated.[12]

Hubbard described the frustrations of his team's efforts. Each night the barge leaked new oil, which his boats pursued the next day. As the days progressed, oil mixed with water to form a mousse-like mixture that fouled even the most efficient skimmers. Waves pushed water, then oil, then water again into the pumps. The skimmer pumps worked less and less regularly. Oil became so thick it clogged the skimmer belts.

As the weather deteriorated into higher seas and bigger waves, these side effects became more uncontrollable. Skimming operations ceased altogether on Day 4, 5 and 6 of the spill because of danger to the crews. After the barge was sunk on Day 9, the skimmer operation halted again in the second week when weather conditions worsened.

Skimmers and booms develop problems in high seas, where waves of six to 10 feet wash over the booms, flooding out oil. Boats holding the booms are unable to remain in position in heavy seas. Oil and water mix together too thoroughly for the oil to stick to the skimmer belt. These problems occurred in *Berman*. Heavy seas and high waves rendered some skimming inefficient because the oil/water mix splashed over the top of the inflated boom and escaped the skimming operation. Even when the ocean was calm enough for skimming, workers encountered another basic problem. The heavy oil began to submerge as it traveled away from the barge, making it difficult to spot from the surface. Hubbard flew above the spill scene to scout for oil for his surface crews. Rough seas churned the waters so severely that waves almost hid the oil from those working skimmers on the surface. Said Hubbard:

> *"It was hard to see the oil on the surface even from the helicopter. My men on the surface couldn't see the discharge at all. I'd talk to my people on the radio. I'd tell them 'come left' or 'go right' in order to intercept the streams and pockets of floating oil."*[13]

Near-shore skimming operations went more smoothly because of increased effectiveness due to better access directly from beaches, and calmer conditions. For skimmers operating in shallower waters near the reef line, the GST used two boats hauling a boom that trapped oil in the belt skimmer. In other sites, Hubbard positioned his people onshore, dropped the 400-pound skimmers by crane from the beach into the water near an oil pocket, fixed the boom into a triangular configuration, and hooked the hose directly from the skimmer into trucks waiting in the Hilton Hotel parking lot. Hubbard monitored each skimming operation daily and reported to the FOSC. He recalled:

> *"My duties included purchasing and costing the equipment, arranging logistics for the equipment to be deployed on-scene, operating the skimmers, supervising the contractors, conducting overflights, and helping with the GST people on board the barge. Quite simply, I was in overload."*[14]

RAMP UP

Private teams battled the spill alongside U.S. government forces. National Response Corporation (NRC) from New York was the worst-case scenario OSRO hired by the RP to handle the cleanup in the first instance, under the supervision of Maritime Bureau, Inc. (MBI), a response management service provider. The insurer for the RP brought in MBI in the early days after the incident. When the RP's $10 million coverage was used up, the USCG kept NRC on scene until January 28 because of the company's prior experience, their good performance record, and the need to maintain continuity in the response effort. Mark Miller, president of NRC, relates how his company began, evolving into one of the finest spill responders in the U.S.:

> *"My family used to run a fleet of fishing vessels. The fishing business wasn't so great. One day we were returning home and learned that there'd been an oil spill in the bay. My dad, brothers, and I rushed home to the farm, loaded up our trucks with straw, and used it to soak up the oil at the scene of the spill. That's how we got started in a business in which today we are regarded as the number one spill response corporation in the U.S."*[15]

NRC maintains a bilingual regional manager in Puerto Rico. When the emergency spill call from *Berman* owners New England Marine Services came into NRC, Miller assembled his "battle staff" in their "situation room" for a full briefing by their regional manager on exactly what was happening at the spill site. Based on the regional manager's on-the-scene report, they determined that *Berman* was a spill of national significance (SONS) that required sending people and supplies at once.

Miller obtained the RP's approval to send seven NRC personnel to Puerto Rico to activate the spill response. His logistics team in Calverton, NY shipped equipment to Puerto Rico, where his ground team trucked equipment from the airport to the sea and began the response in what Miller termed the "push and pull" of planning.

Because of the USCG and insurance underwriter's prior experience with the RP, New England Marine Services, involving a spill of 10,000 gallons of No. 6 fuel oil in San Juan Harbor from the Bunker Group's *BGI Trader* just three weeks earlier, all players assumed their positions quickly. While there was a duplication of effort between RP troops and USCG until Day 8 (January 14), operations proceeded smoothly. MBI and USCG took lead oversight roles. NRC and Crowley, along with USCG, USN and GST members, assumed ground and sea operation responsibility.

Using a *spill specific cascade schedule*, NRC feeds data into a computer to determine where the nearest equipment is, which contractor has the resources to meet the response needs, and how best to transport inventory and people to the scene – the "ramp up" phase, to use Miller's term.

NRC's spill cascade system worked well in *Berman*. Within 20 hours of notification, NRC accumulated, sent and delivered the first round of massive equipment to the staging area, including a full communications setup; 19 hours passed between the second round request and delivery of the cargo; within 22 hours from the Coast Guard's demand, a third round of airlifting was completed; and the final fourth round took only 14 hours from the Coast Guard's last request.[16]

Miller and his communication staff arrived at the spill site January 8. Miller met with Stanton daily. NRC's primary areas of response were fighting the free-floating oil in the initial cleanup stages with their skimming and booming equipment, and then conducting beach cleanup, particularly in the Caribe Hilton Lagoon and near the stadium area. NRC arranged transport of oil collected on Crowley's *Barge 250-1* to Sun Oil Refinery.

By January 28, Miller considered NRC's role as primary contractor at an end. "Why wear out our welcome?" he said. He felt NRC's continued presence was a service made redundant and handled quite efficiently by other agencies and federal personnel. Miller sounded elated with NRC's role in the response:

> *"We blew away every guideline in the regulations. We exceeded what was required eight to 10 times. Boom and skimmers were deployed in remarkably quick time. My local contractors were pre-staged and ready to go, with lots of equipment on scene. We airlifted the rest rapidly. Puerto Rico is not a high volume port. Under OPA, technically we have 24 hours to get everything into place. In* Berman*, the first boom was splashing down in the water within hours of the spill."[17]*

"PUTTING IT ON THE BUBBLE"

The second major priority of the early spill response required securing the oil source by removing the barge. LCDR Hooper coordinated this phase of operations. Hooper devotes about half of his time to oil pollution response and ship salvage. With the help of Kevin Peters and others, Hooper quickly sized up the situation. The vessel remained aground, spilling oil and littering the beach. To stop the oil, SUPSALV had to control the *Berman*.

Three basic salvage options were available for the damaged vessel:

- Seal the barge in place and leave it as a permanent fixture off Punta Escambrón.
- Remove the barge permanently for disposal at sea.
- Haul it to another port.

The barge was too badly damaged and the seas too rough to consider any repair of the vessel at the grounding site. Hauling the barge into San Juan Harbor was far too dangerous. San Juan Harbor is reportedly one of the roughest ports of entry in the Caribbean. If the huge barge broke up at the mouth of the harbor, which was a distinct possibility, it could block the channel, effectively closing Puerto Rico's major port to all incoming cargo, including food, and cause economic disaster for the 3.8 million Puerto Ricans resident on the island and their millions of visitors.

Even if the barge made it successfully through the mouth of the harbor, the port had insufficient facilities to repair her. Equally unappealing was the idea of towing the leaking wreck into the Dominican Republic or one of the U.S. Virgin Islands. Understandably, tourist-dependent islands or countries are not receptive to an oil-gushing barge moving through their clean waters and pristine beaches.

The only real choice was to remove the barge from the reef and sink her.

According to SUPSALV, the RP's salvage plan was seriously flawed and the FOSC rejected it. Among other problems, the RP's salvage plan contained no destination for the barge once it was refloated, no cleared receiving port, and no arranged burial ground. SUPSALV produced an acceptable proposal for sinking the ship. While we commonly think of salvage as recovering a damaged vessel, in other cases, like *Berman*, salvage includes the option of sinking a vessel.

Two days of preparation to accomplish this objective began with lightering as much remaining oil as possible out of all the breached and unbreached tanks and onto the *BGI Trader*. Then salvage workers pressurized the barge by sealing shut all topside holes, valves and vents where air or water might rush into or out of the vessel. These included openings on top of the deck where fuel was loaded or off-loaded, tank vents, and the tears in the hull. They could not seal the ruptured bottom, but that was not a problem because air forced inside would bubble to the top, lifting the ship off the reef where it could be hauled out to sea – "putting it on the bubble," in SUPSALV lingo.

SUPSALV performed pressurization tests on the damaged tanks January 14 to make sure they had correctly calculated the air pressure needed to keep the barge afloat. Removal operations began at 7:30 a.m. January 15. Personnel from DonJon Marine, a company whose business includes sinking ships, hooked a tug to the barge, and SUPSALV pressurized the tanks. About 11:45 a.m. the damaged

High seas complicated response efforts during *Berman*. (Reprinted by permission of U.S. Coast Guard MSO, San Juan.)

barge slowly rose off the devastated reef to cheers from observers.[18]

Well before that moment, however, a decision had to be made: where in the world to *take* it?

PUSHING THE SAFETY ENVELOPE

The only junkyard for ships and unwanted explosive waste available to the FOSC in the Puerto Rico area was the Defense Dumping Grounds, about 20 nautical miles off the northeastern coast of the island. This was a potentially suitable burial ground for the *Berman*. Despite the USCG's best efforts, a substantial amount of oil remained on board the barge because the oil viscosity became so great that the pumps simply could no longer push the oil through the hose to the lightering vessel. As the increasing risk of worsening seasonal winds threatened to break the barge apart near shore, removal became necessary. National Oceanic and Atmospheric Administration (NOAA) scientists studied the options in the days preceding the actual sinking, and advised Capt. Ross that cold temperatures in the deep waters of the Puerto Rican Trench would cause the remaining oil to congeal in the sunken barge and prevent it from flowing into the ocean.

SUPSALV had several options which included: (1) towing the barge to the Navy's Munitions Dump and sinking her in about 6,600 feet of deep water; (2) transporting the *Morris J. Berman* to a deeper burial site 30 nautical miles further offshore with depths of 4,000 meters (8,000 feet); or (3) traveling 100 nautical miles off the northern shore of Puerto Rico to depths of 8,000 meters (12,000 feet.) The FOSC chose the dumping grounds.

Detractors criticized the final choice. The FOSC responded with convincing arguments. NOAA forecast two days of calm followed by the roughest weather experienced since the start of the spill. Time was running out, with the increasing likelihood that the barge might break up and spill more oil close to shore.

As water depth increases, temperatures drop. NOAA advisors speculated that temperatures would be low enough at depths of 6,600 feet to keep the oil below its *pour point* (40 degrees F.), the temperature at which petroleum solidifies from its liquid form and is less likely to rise to the surface. Technical advisors believed that the water temperature at 12,000 feet would not be significantly below this pour point. Slow releases were predicted to occur at any of these depths. [19]

There were significant drawbacks to reaching deeper waters. The tug towing the barge would have to haul it even farther north beyond the dumping grounds, increasing the risk of free floating oil impacting the nearby Dominican Republic. Even with the momentary calm seas, the waters were rough given the fragile state of the barge. The FOSC was concerned about losing Gulf Strike Team members and private contractors off the barge or endangering the tug crew who were hauling an unstable vessel on a towline in rough ocean. No one was excited about the prospect of moving a heavy, unsafe barge farther out to sea in the dark. In their ardor to beat the spill, those involved with the salvage operation were already pushing the safety envelope. The decision was made to push no further. The sooner the salvage crew and equipment were off the barge, the better.[20]

During the 20-mile transit to the Defense Dumping Grounds, Hooper removed his men and machinery from the barge by helicopter. He and five crewmen on board closed the tank air valves, checked the air pressure, and were finally lifted off.

Dale Springer was the last man off the *Morris J. Berman* before she slipped to the bottom of the ocean. Since his 18th birthday, the 54-year-old salvage master had spent his life around the sea. He worked for DonJon Marine, and treated this final responsibility with the same matter-of-fact professional manner he applied to all his work. Springer opened the valves on each of the tanks and gingerly stepped off the barge onto the tug which had pulled her out to sea.

One eyewitness to the sinking, mariner A. Carlo Calanni from Marine Spill Response Corporation, said the sight moved him to tears. Calanni said the *Berman* "sounded like a screaming woman" as she slowly sank beneath the waves. He thought of her wailing, "Why me? What did I do to deserve this fate?"[21] At 6:35 p.m. the USCG reported the *Morris J. Berman* had sunk.[22]

Hooper characterized the salvage effort:

> *"This spill was an 'average job,' but with high visibility. We were lucky in having Capt. Ross in charge. His instinctive response to managing the spill was correct. He was attuned to what needed to be done. Under the circumstances, the FOSC handled this operation admirably."*[23]

MAJOR RELEASE NUMBER TWO

Sinking the *Morris J. Berman* should have ended the oil flow, but it did not. Capt. Ross explained that there was no question some oil would enter the marine environment during the sinking operation. Responders at the spill site had the job of recovering as much oil as possible as quickly as possible, limiting the amount leaking into the waters. The *Caribbean Responder* skimmer with two tugs pulling booms followed behind to mop up any spilled oil, as did other skimmers. Following the barge out to sea from his 110-foot Coast Guard cutter, Stanton saw "the 'clingage' washed out of the barge as the waves hit the *Morris J. Berman,* and geysers shot out of the holes in the bottom and sides, leaving a trail of oil behind."[24]

Detractors fault the scientists for not correctly predicting the extent of the subsequent oil release. Oil did *not* gel inside the barge, but instead leaked out in a second major release that traveled west along Puerto Rico's entire northern coast, threatening sensitive ecological areas. Capt. Ross answered with "an emotionally difficult" fact to accept, that he "couldn't do the impossible." At the time of sinking the barge, there was nothing further the USCG, SUPSALV, or anyone else could do to prevent oil from entering the environment – "Mother Nature was in charge."[25] An additional 200,000 gallons of No. 6 fuel oil flowed from the barge, either from its transport to the dumping grounds or in the days and months following the sinking. A large slick appeared over the sunken vessel to form a floating tombstone.

Wind and ocean currents drove this oil slick toward pristine and sensitive environmental areas on beaches almost 40 miles west of the dumping grounds, along a stretch of 12 miles of coastline between the Puerto Rican towns of Isabella and Borinquen. Tarballs and belts of tar stuck in the trough lines along Aguadilla, Piñones, Jobos and Shaks Beach, this last a vulnerable home to manatee caves. The additional discharge triggered another spill cleanup far from the grounding site.

As predicted, bad weather arrived January 17 and prevented skimming for several days after the sinking. SUPSALV resumed offshore skimming efforts January 24-27, and then packed up the bulk of their equipment and shipped it home. Hankins decontaminated his equipment and shipped it back to the U.S., with the last equipment returning

INCIDENT COSTS AND RESPONSE EFFECTIVENESS COMPARISONS

Spill	Cleanup Costs[1] $/Gallon Spilled	Recovery[2] Effectiveness	Damage Costs[3] $/Gallon Spilled
EXXON VALDEZ[4]	$285+	Unknown – Generally Assessed as Low	$1,000 +/-
Tampa Bay[5]	$206+	Unknown – Estimated at approximately 80%[6]	$91+[7] (partial)
BERMAN	$104[8]	78%[9]	$17[10] (projected total)

Overall BERMAN Effectiveness, including lightering – 88%+[11]

Notes:
1. Does not include lightering and/or salvage costs or any damage costs.
2. Does not include oil saved through lightering and/or salvage.
3. Damage Costs including Third Party economic and property damages as well as Natural Resource Damages.
4. EXXON VALDEZ response cost and damage figures taken from Financial Cost of Oil Spills, Dagmar Schmidt Etkin, Oil Spill Intelligence Report and Cutter Information Services. (1994 dollars)
5. Cost and damage figures for the Tampa Bay spill are based on various sources close to this response, including response contractors. Because of pending litigation, finalized figures have not been released and caution is warranted in using these figures.
6. FOSC Tampa figures.
7. Includes $10M in Third Party property and economic damages and a partial NRDA figure of $20M. (Figures approximate.) Final mitigation activities and NRDA still pend at Eleanor Island, an inland mangrove stand and bird sanctuary. The final Natural Resource Damage figure will likely climb.
8. OSLTF and WQIS costs included. Government monitoring/oversight and public safety costs included.
9. Based only on Sun Oil, Puerto Rico figure for recycled liquid oil. No landfilled oil included.
10. Projected total Third Party economic and property damages and Natural Resource Damages. Includes claims processing costs.
11. Based on recycled liquid oil recovered through lightering, skimming, and submerged oil recovery.

("The Response to the T/B Morris J. Berman Major Oil Spill." Reprinted by permission of U.S. Coast Guard MSO, San Juan.)

February 1. Hankins remained on the scene until the recovery was con-cluded.[26] Hubbard returned to his base with the bulk of the force's equip-ment. Other team members returned to San Juan when the towing and sinking of the barge resulted in its second major release. While the off-shore skimming operations were over for all intents and purposes, the battle lines shifted to recovering oil washed up on the beaches and rocky shorelines, or submerged in lagoons and shallows.

OIL BUDGET: COSTS OF RESPONSE

According to the most reliable figures available, of the total USCG *Berman*-related expenditures of *$81,700,000*, salvage and lightering cost *$4,052,000* and offshore skimming another *$2,049,000*.[27] The balance of the USCG's funds were spent as shown in the accompanying chart.

Were these funds well spent? Members of the international shipping community criticized the *Berman* spill response, particularly the offshore operations. Joe Nichols of the International Tanker Owner's Pollution Federation, Ltd. (ITOPF) suggested two serious flaws in the response operation in his critique of the FOSC's actions: (1) the FOSC had far too much skimming equipment on scene for the circumstances, particularly given the minimum required in the Area Contingency Plan (ACP); and (2) the skimming operations were highly ineffective given the volume of oil recovered. Says Nichols:

> *"The average recovery rate achieved over this period was about 137 barrels per day, which represents a skimming efficiency of about 0.1 percent. The total quantity of oil recovered from the water was estimat-ed to be 2,700 barrels, which represents about 20 percent of the spill vol-ume. However, it is believed that the bulk of this was collected via the shoreline where the oil had been compressed by onshore winds."*[28]

POLLUTER PAYS

The U.S. is engaged in a major dispute with European and other inter-national communities over how oil spills should be managed and who should bear the cost. The argument for both sides is discussed in greater detail in Section III. This is the dispute in brief:

The primary thrust of OPA 90 (the primary U.S. regulation regarding oil spills) is that "the polluter pays." Whoever caused the spill is ultimate-ly responsible for all costs, from cleanup to natural resource damages. This position ensures that owners and operators in the industry clean up their act and attend to the business of making their industry safe. Without this economic incentive, the U.S. believes that outrageous and environmental-ly damaging spills would continue to occur. As one commentator put it, the U.S. refuses to wait around for "the spill of the month."

The insurance community opposes the scope of damages covered by OPA 90, and the potential of unlimited liability under OPA 90. Under international treaties (which the U.S. refuses to sign) they have achieved a limit to what they will pay for based on an industry standard of a dollar amount per gross ton for the vessel involved.

The international attitude of *least cost response* trickles down through every phase of the cleanup. Emphasis is placed more on shoreline defense and cleanup than on mechanical recovery of oil at sea using skimmers and booms. Natural resource damages are not compensated. OPA 90 embraces the opposite view. The U.S. requires compensation for environmental damages of all kinds.

According to Capt. Ross, an ITOPF representative who visited him in May 1994 stated, "There is no historical, legal or moral justification for the idea that those who have been damaged by an oil spill should be entitled to compensation for those damages."[29]

The first and best response considered by European and other world communities is the *do nothing* approach: let Mother Nature take her course. Non-U.S. responders more readily adopt the concept of *sacrificing* a particular beach or natural resource. In their view, it may be better to allow the oil to saturate one particular area and then focus the cleanup effort on those beaches once spilled oil hits the shoreline. The foreign communities believe this methodology is far wiser than spending exorbitant sums trying to clean up all recoverable oil from the ocean with often ineffective skimming techniques before the petroleum hits the beaches.

To say that Capt. Ross takes exception to statements like those made by ITOPF officials is an understatement, and his rebuttals are pointed and persuasive. Addressing criticism that *Berman* employed too many skimmers with too little effect, Ross explained that he ordered the skimmers he believed were needed after considering many factors:

- Volume of oil recovery capacity.
- Weather.
- Suitability of the oil to skimming procedures.
- Conditions under which skimming would be conducted, in sheltered water, in open ocean, and in semi-sheltered waters.
- Sea conditions near the grounding.
- Close proximity of the grounding site to the San Antonio Channel.
- Substantial pockets of oil trapped near the shoreline during the response.
- Mechanical breakdowns.

Ross admits that the skimming operation was not flawless. For example, the 208-foot *Caribbean Responder* was employed in conditions unsuitable to

THE *MORRIS J. BERMAN* SPILL OIL BUDGET

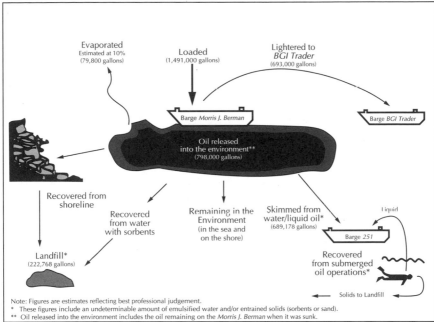

Evaporated
Estimated at 10%
(79,800 gallons)

Loaded
(1,491,000 gallons)

Lightered to
BGI Trader
(693,000 gallons)

Barge *Morris J. Berman*

Barge *BGI Trader*

Oil released
into the environment**
(798,000 gallons)

Recovered from
shoreline

Recovered
from water
with sorbents

Remaining in the
Environment
(in the sea and
on the shore)

Skimmed from
water/liquid oil*
(689,178 gallons)

Liquid

Barge *251*

Landfill*
(222,768 gallons)

Recovered
from submerged
oil operations*

Solids to Landfill

Note: Figures are estimates reflecting best professional judgement.
* These figures include an undeterminable amount of emulsified water and/or entrained solids (sorbents or sand).
** Oil released into the environment includes the oil remaining on the *Morris J. Berman* when it was sunk.

(Source: "The Response to the T/B *Morris J. Berman* Major Oil Spill, San Juan, Puerto Rico, USCG. Reprinted by permission of U.S. Cost Guard MSO, San Juan.)

its best use, such as requiring it to run parallel to shore in eight to 10-foot seas with winds of 18 to 20 knots and more. MSRC, the owners, learned much about how best to operate this equipment and have made substantial changes in their operations as a result.[30] The Coast Guard's VOSS skimming equipment was overwhelmed by offshore swells which caused the outrigger arm to lift out of the sea, making skimming less effective.[31]

On the whole, skimming proved effective and the equipment used was reasonably necessary. Capt. Ross regards the figures used by Nichols in his critique as being totally "invented," and counters with fact. According to industry figures from Sun Oil Refinery, the actual amount of oil recovered from skimming and submerged oil operations and recirculated through their refining process was 16,409 barrels (689,178 gallons), of which 90 percent was oil. Sun Oil in Yabucoa, PR treated an additional 10,000 barrels of a predominantly water-oil mix; the actual amount of No. 6 fuel oil recaptured in this process is unknown. Finally, an unspecified amount of oil sludge was mucked out of the bottom of the barges and deposited in landfill. Skimming and other operations netted recovery from the environment of more than 16,400 barrels or *698,000* gallons of oil. Further, in the overall context of *Berman* costs, the per gallon cost for offshore recovery was about one half the cost for shoreline recovery, the response effort more often favored by ITOPF. Considering the potentially catastrophic nature of the *Berman* discharge, this must be regarded as a triumph.[32]

BODY COUNT PHENOMENON

Capt. Ross tends to scoff at the heavy reliance Nichols and others place on these numbers. He recalled the distasteful practice by U.S. military forces in the Vietnam War of reporting sensational enemy body counts to suggest victory was within our grasp. Ultimately, like that counting of coup, recovery numbers are arguable, difficult to prove either way, and sometimes meaningless.

There is no scientific precision when it comes to how much oil is recovered by skimming and salvage operations. These numbers are only an indication of efficiency. U.S. oil spill response concerns are two-fold: that cleanup be accomplished as cheaply as possible, but not at the expense of performing an effective response. A reality check is needed. What did the impacted area look like before the spill? During the spill? After the cleanup? What damages were avoided or minimized by response operations?

Two years after *Berman*, I walked the cleaned beaches and examined the rocky shoreline and riprap (the man-made rocky embankment preferred by some resort hotels) affected by the oil spill. There is still *Berman* oil present, but in much reduced quantities. I can testify that most of the visible oil is gone. Regardless of man's best efforts, we simply cannot remove all spilled oil from the environment. That is why preventive measures to avoid spills are so critical.

Capt. Ross holds a clear opinion about the overall effectiveness of the *Berman* spill response:

> *"Well, of course, I'm more than a little biased. But I look at the figures as a relative indicator of the oil recovered and I use my eyes to see the physical evidence of how the beaches and affected areas are now. Based on this, I would say we had an extremely effective response. Part of this was luck and the other part was good cooperation among those involved and a common sense approach by the commonwealth. We reacted with great speed. We were fortunate to have a narrow shelf shoreline in a high energy environment. This meant whatever we didn't recover from skimming the surface or from the bottom came onto the beaches where we were able to clean it fairly well."*

Capt. Ross's parting shot echoes almost every spill responder: "In the oil recovery business, it's entirely possible to do everything right and still fail. I don't believe that's the case in *Berman*. The fact of the matter is that the *only perfect oil spill response is prevention*."[33]

RESCUING THE LIVING KINGDOM

HABITAT DISRUPTION

Scientists resort to euphemism, lump together all deaths, and call it *habitat destruction* when they do not know how to place a monetary value on wildlife destroyed by man-made disasters such as oil spills.

The two *Berman* releases spilled oil over 30 miles of coastline and 20 square miles of sea surface. To say the least, the habitat of all living creatures within that radius was *disrupted*. Fish, corals, sponges, gorgonians, algae, seagrass, sea urchins, birds, turtles, manatees, shellfish and other species experienced the effects of the oil in all its forms: liquid, tarmats, tarballs, pancakes and slicks.

In the first weeks after the spill, those charged with aiding injured wildlife collected 5,687 oil-affected organisms. Of these, at least 5,268 died due to intoxication or suffocation from the oil. Antonio Mignucci-Giannoni, the scientific coordinator and chief author of the preliminary mortality assessment conducted of the spill and its effect on marine creatures, believes these figures represent only about 10 percent to 15 percent of all creatures affected. The *echinoderms* (sea urchins and sea stars particularly) were hit hardest, making up 62 percent of all reported deaths. Of those animals not killed immediately that were brought in for rehabilitation, their recovery rate was less than 50 percent. "In general, considering all spills and all species, oiled animals brought in for rehabilitation showed a 44 percent probability of survival."[1]

Rare and endangered species did not escape *Berman's* reach. The spill hit endangered sea turtles, sandwich terns, brown pelicans, an Audubon shearwater, a dolphin, brown boobies, and many others. Antillean manatees and a humpback whale in the vicinity escaped without harm.[2]

FRESH RUBBLE

James Timber of the Puerto Rican Dept. of Natural and Environmental Resources (DNR) studied the barge's effect on the reef spill site. Many use the term "dead reef" when referring to the *eolianitic* reef at Escambrón Point. This term is completely inaccurate. Although an eolianite reef is not formed in the normal manner by hard coral skeletons, it does house a full system of living organisms. Essentially, this structure is a cemented sand dune, with both hard and soft corals, sea fans, and an active benthic and bottom community growing over the rock surface and living in the sediment around the floor of the structure. In comparison to our short lives, the Escambrón reef is reportedly about 7,000 years old. Juvenile fish make their home there, as do stony corals in small colonies of less than 45 centimeters, sponges, small urchins, sea anemones, and many other coral community members.

That was the case before *Berman*. James Timber and two Navy Seals dived the reef attempting to see what the barge had done to this living community. They followed a polypro line down to the reef because of rough seas. Timber and the others carried cameras, an underwater slate and measuring tape. Waves thrashed around them as they dived among the rubble left by the barge. "It was not a pleasure dive. Conditions were choppy and miserable," Timber recalled.

The dive crew compared the grounding site, an area about 300 feet long by 90 feet wide, with other parts of the reef that stretched for miles and ranged in depth from eight to 25 feet near the scene of oil discharge. They swam in organized transects across the reef until they noticed any significant difference in the composition of the benthic and other communities, and then photographed sections of the reef, took measurements and mapped the various areas for types of damage: partial, complete, and rubble.

Timber recalls what they saw:

> *"In the area where the ship ran aground there was considerable impact. There was complete destruction from the fore reef to most of the back. The reef looked like fresh rubble. In the middle area, where the ship didn't grind into the reef, there were big depressions, from rubble being thrown onto the coral and fans. After the barge was removed, rubble pieces 20 to 30 centimeters wide were thrown around. Without doubt, these medium-sized pieces will get cemented into the reef and modify the environment.*
>
> *"Many of the reef creatures were destroyed completely, especially the benthic ones, from the scraping and crunching. Others were partially destroyed, such as the gorgonians attached to the substrate. Corals were torn apart. Other animals lost their cover when the rubble cemented over their hiding spots. What will happen to them remains to be seen."*[3]

Timber considered how long it might take for the reef to recover, and what, if anything, DNR might do to aid the process.

> *"We'll leave the reef alone for a time and see what happens. We have to control the water quality and make sure no chemicals are*

dumped along the shoreline. After a time we could implant new colonies of soft or hard corals, if we got the go-ahead to do so."[4]

Puerto Rico awaits the outcome of a lawsuit the commonwealth filed against various *Berman* RPs. Further reef rehabilitation depends on the suit and damage assessment.

Vance Vicente found damaged coral communities along the shoreline that were similar to those at the grounding site. Oil rots the bases of sponges, killing them quickly. In many places along the affected coast, not one single species of sponge was found. In other sites along the spill line, Vance and other observers failed to see a single species of live gorgonians. The macrophatic algae suffered a direct hit. These algae make up the base of the food chain for the reef community. Much of the alga was damaged extensively and appeared to be in various states of decomposition in the early days after the spill. Vicente saw "tufts" on much of the coral he observed, a sign of substantial damage to the living coral, and expected further mass mortality. Post-spill coral cover was less than 1 percent outside the Caribe Hilton Beach.

This was not surprising, given Vicente's description of the oil present in the area:

"Thick mats of oil were seen to form blobs which were 'dripping towards the surface,' forming gorgonian-like branches. On the other hand, tarballs from about .5 cm to 2 cm in diameter were found commonly either between the ripple marks, or over the flat sandy bottom."[5]

Vicente produced daily reports about what creatures were affected and where, and assessed the oil impact on animals and flora resident in the sandy beaches, muddy bottoms, coral reefs, rocky shorelines and lagoons of the spill sites. He collected specimens from all affected areas, dived the Escambrón reef, and walked the beaches and rocky shorelines, recording the destruction wrought by the

Oil sheen approaches coastline. (Reprinted by permission of U.S. Coast Guard MSO, San Juan.)

oil on these once-living communities. According to one source, after discovering more than 500 dead sea urchins, Vicente simply threw up his hands and stopped counting.[6]

STERILIZED

Fish, sea urchins and other invertebrates in the sandy intertidal, rocky intertidal, and benthic zones in the path of the oil fared no better than the corals. There is no escape for these creatures. Once touched by oil, they are smothered or asphyxiated by the substance. Workers could do little but pick them up, bag them for later evidence of damage, and place their remains in freezer storage lockers. Geologist Felix Lopez, a member of the U.S. Fish and Wildlife Service (USFWS) ecological team that helped NOAA's scientific support coordinator, described conditions around the Condado area immediately adjacent to the spill:

> *"These bottom marine organisms were literally 'sterilized.' There were so many dead, we stopped counting. When you see a figure like 3,000 dead urchins, just assume this is about 10 percent of the total killed by the oil."[7]*

Fish kills were significant in the first days, including highly popular restaurant and dinner table fare such as mackerel, silk snapper and yellowtail snapper. One observer reported thousands of fish going belly up in the spill site area. Normally, fish will sense oil and swim away. Reef fish, however, rarely travel far from their reef habitat. Many fish who lived on the reef near Punta Escambrón died immediately. Others suffered skin ulcers, partially rotted fins, blindness, tissue rot, missing or damaged eyes, spines overgrown with alga, and a myriad of nightmarish injuries. Eventually responders hauled away the dead reef inhabitants that washed ashore with the other waste.[8]

MYSTERY SPILLS

As if *Berman* and its environmental destruction were not bad enough, certain opportunists took advantage of the spill and made matters worse. Beginning January 21 and continuing until February 11, ominous reports surfaced of oiling on other Caribbean islands and parts of Puerto Rico far removed from the spill site, including the shorelines of Culebra, St. Croix, St. John, St. Thomas, the east coast of Puerto Rico, Vieques, and several small keys. At first responders thought this was more *Berman* oil, but Coast Guard lab analyses revealed that *Berman* was not the source of these additional discharges.

Deliberately dumping oil at sea within coastal waters is a criminal offense and a violation of treaty in international waters, but there are always those who will break the law to serve their own needs, such as saving the expense of coming into port and cleaning out their bilges properly. Figuring a little more oil would not be noticed in all the confusion, tankers, barges and other vessels near Puerto Rico simply opened their bilges and pumped oil into the sea. Others washed their tanks of petroleum and let the runoff flow into the ocean, again as a way of saving cost.

These *spills of opportunity*, also known as *mystery spills* because of their unknown source, complicated the recovery and cleanup efforts by expanding the affected spill area and further threatened sensitive turtle nesting and sea bird habitat. Shaking his head in dismay, Felix Lopez reflected on the human behavior that condones mystery spills:

> *"Whenever there's a major oil spill, people take advantage. There was a spill in St. Kitts two to three years ago right before Easter. All I can tell you is that people took that opportunity to dump oil into the ocean. That seems to be the normal pattern."*[9]

RED CARIBEÑA VARAMIENTOS

The initial outlook was bleak for wildlife affected and injured by the oil. But a local group of volunteers and a small society responded to the disaster with open hearts and hard work. Antonio Mignucci-Giannoni, a local Puerto Rican scientist, and longtime resident and teacher identified only as Prof. Williams, along with others, created the Caribbean Stranding Network (CSN) in 1989 to protect threatened and endangered species. Their most publicized success was the rescue and subsequent release of Moises, one of the few remaining West Indian manatees in the Caribbean. Their network, also known as *Red Caribeña Varamientos*, has grown from a handful of enthusiasts to 250 or more members throughout the Caribbean.

Within hours of notice of the *Berman* spill, Mignucci and his small crew arrived on scene. DNR contracted with them to rescue and rehabilitate live organisms, and to collect specimens for a future damage study. Mignucci's first impression: "It was horrendous. For every animal in the oil's path, the situation was bad."[10]

Mignucci and 20 to 30 volunteers worked for "what seemed like 24 hours a day for weeks" to count, identify, number, tag and freeze dead specimens and to collect and treat the survivors. Frozen specimens were stored at the Recinte at La Parguera, a research facility associated with the University of Puerto Rico, piled to the ceiling in freezers. The disturbing sight contrasted sharply with the harmonious memories of past encounters with live marine creatures in their untainted habitats.

Cleaning an oiled bird or sea animal is a time-consuming labor of love. First the veterinarian on staff or another qualified person thoroughly examines the animal, looking for and noting signs of stress, the degree of oiling, and the animal's overall condition. Based on this assessment, the worker either places the creature in a dark, warm box to calm it, or begins treatment. Care workers feed the animals olive oil to aid the intestinal lining and provide relief from petroleum toxicity. Staff administer a non-steroid antibiotic on the cornea of each eye before cleaning begins.

Stabilized individuals are sent to a cleaning station for bathing. The primary solvent used is olive oil, applied by towel and massaged into the animal's plumage, carapace or skin. CSN workers then carefully restrain the birds, turtles and others, and immerse them in tubs of warm water with mild detergent for more massage. Workers must be careful not to damage the waterproofing of the feathers or to immerse the heads of any of the animals.

At the third treatment station, animals are rinsed and dried by forced air heat dryers and absorbent toweling. This process is accomplished quickly to avoid chilling. Depending on the animal's health and stress level, cleaning is repeated 24 to 48 hours later until all oil is removed.

Like patients in an Intensive Care Unit, treated animals are observed closely and fluid therapy is administered when necessary. Each bird is housed in an isolated, warm cardboard box, draped with a towel. The boxes rest on heating pads. Recovering birds are later placed in individual mesh cages.

Volunteers feed the animals a fish gruel mixture by using a lubricated stomach tube and syringe. When they are able to handle more food, the birds or mammals receive small sardines or a hand-fed fish gruel mix. This specialized process requires considerable care, patience, time and energy. The fortunate animals who recover are released. Birds are transferred to an enclosed flying cage so they can dry their wings, exercise, feed and dive on their own. Birds and sea turtles are finally tagged before release in a "safe," oil-free location.[11] Even crabs were cleaned by hand, on site, with cotton swabs and soft soap, then released in uncontaminated spots.

Puerto Rican residents assisted by calling the CSN's toll-free number whenever they found injured or oiled wildlife. Volunteers manning the phones instructed residents where to take injured creatures. In the first days, the phones rang constantly as residents roamed the beaches trying to find and rescue maimed and oiled animals.

While the CSN did an exemplary job and managed to save many of those animals rescued, objective observers question what would have happened had the spill affected larger numbers of wildlife. Limited resources and materials hampered their efforts.

Rick Dawson was the Damage Assessment Specialist assigned by the USFWS to the *Berman* spill. Dawson has been in the oil fighting business for years, and participated extensively in the *Exxon Valdez* spill wildlife assessment. His primary job was to assist the FOSC at the site and to over-see the response so as to mitigate damage to the environment and natural resources. Dawson's additional duties included accumulating data on injuries sustained by the natural resources, birds, fish, invertebrate crea-tures, reefs, mangroves, sea turtles, seagrasses, and other life affected by the oil. His assessment of the Caribbean Stranding Network's performance:

> *"They did a great job with limited resources. CSN was inade-quately staffed and equipped to handle more than a few dozen animals. I had two contractors on retainer in Tampa with full field set-ups and 11 vets on staff. If there had been any more birds or mammals involved, my next call was to CDR Stanton to request a C-130 to load every-thing up in Tampa and bring it all back to Puerto Rico. The CSN was more like a 'backyard' operation. Fortunately, from the USFWS view-point, we were very, very lucky. Fish kill was localized. Sea bird rook-eries appeared not to be directly impacted. Sea turtles seemed to 'duck' the oil. There's not much you can do for echinoderms."*[12]

According to Dawson, spill responders in *Exxon Valdez* spent as much as $85,000 per sea otter to save maybe 30 percent of some 18,000 affected otters. Some question the efficiency of such excessive front-end costs, since funds spent on rescue are not available for much needed rehabilita-tion and restoration projects after the oil is removed. Dawson saw the same problem with wildlife response in *Berman*. In his opinion, too much time and effort was spent rescuing echinoderms, which may be consid-ered of limited commercial or other value unless they are rare or endan-gered. He believes the better approach would have been to limit front-end recovery expenditures, and throw those resources into upgrading CSN's equipment and manpower base for future effective use.[13]

ONE DROP

Not all animals washed up onto shores and beaches. The spill threat-ened diving birds, gulls and terns, wading birds, shorebirds, raptors and waterfowl, including rare and endangered species who winter in or inhabit the region year-round. Of most concern are diving birds and waterfowl who become oiled from diving into polluted water or floating on oil slicks. Shorebirds do not usually come in contact with oil unless the substance sticks to their feet, legs or beaks. Raptors and falcons consume oil indirectly by eating oiled prey such as fish or squid.[14]

"One drop of No. 6 can kill turtle or bird eggs," Dawson warns. The responders took this threat seriously. The USFWS started a monitoring program immediately. This project involved shoreline and nesting area surveys, mostly by boat, but also by foot and helicopter. Jorge Saliva, a local Puerto Rican scientist and bird expert, along with other USFWS representatives, conducted daily early morning and afternoon surveys by boat to establish where the birds were found and how much oil they had contacted. The greatest number of birds affected were seabirds, like brown boobies, diving pelicans and the brown gannett. These birds feed offshore in areas similar to or near the Escambrón reef, where they encountered submerged or surface oil.

Saliva flew by helicopter weekly to check islands near the affected Puerto Rican mainland within range of the seabirds, including Cordillera Key, Culebra, Mona and islands on the west coast. Seabirds can fly 150 miles, and may range 15 to 20 miles daily from their nesting areas to feed. Saliva inspected nesting areas high in trees to discover how many birds were oiled and evaluate potential toxic contamination of bird eggs.

USFWS had limited options when they found oiled birds in the rookeries. Saliva said:

> "Many times we can't recover oiled birds. We might try to trap them when they're on buoys, but they'll usually fly off. In the rookeries, we'd be wasting our time trying to catch a bird that's oiled. It's almost impossible to get onto the Cordillera Key, for example. By the time we did get there and then somehow managed to capture the oiled specimen, it would be too late for the bird. So we just observe them and leave them alone."[15]

Saliva uses the Sooty Tern as an example of oil's effect on a seabird. This bird feeds offshore, often travelling 15 to 20 miles out to sea. Because of its waterproof feathers, the tern is light and highly maneuverable. The tern waits for a fish or squid to jump out of the water, then swoops down to catch its prey. If its feathers lose their impermeability, the Sooty Tern may still be able to dive and even to capture the fish, but it won't be able to lift out of the water. The feathers become so soaked with oil and then with water that they lose their lightweight, insulating features. The bird remains stranded in the sea and eventually freezes or perishes from pneumonia.

USFWS will monitor affected rookeries and observe the condition of birds hatching in the seasons following *Berman*. They will study seasons before and after the *Berman* spill to compare the number of eggs hatched, populations of surviving chicks, and how many survivors suffer deformities.

Sea turtles are equally vulnerable to oil injuries. Adult sea turtles can usually detect and elude oil discharges. Unfortunately, some mistake tarmats for sponge, a basic food source which tarmats resemble. Ingestion of this toxic substance is fatal to mature adult sea turtles. While the adult population can suffer harm, the deeper concern is the harmful effect of oil on the thin shell of newly-laid turtle eggs. If oil washed onto beaches seeps into the egg shells, the embryos die. Turtles return to the same beaches year after year to nest, starting in late February. Their routine is fixed and their place of return honed by remarkable instinct. The Aguadilla area contains a number of high use turtle nesting areas. These and other potential turtle-nesting sites along the coastline received high priority attention and cleanup. Saliva described the beach cleanup:

> "We took crews out to the prioritized beaches and hand-picked the oil tarmats and tarballs. We raked some areas, and in others we actually removed the top layer of sand, because turtles dig into the sand to lay their eggs. I felt that we got the sand as clean as we could, given the circumstances."

Saliva considers sea turtles part of his people's heritage. Commonwealth residents used to eat turtles. Today, laws prohibit catching, selling and serving these endangered species, and poaching is rare. Saliva believes that sea turtles have become an even more meaningful natural resource. "Turtles benefit everyone," he says. "Watching one swim is like watching a seabird dive, or a Caribbean sunset. They are part of our life, a value to everyone, part of our heritage."[16]

Three sea turtle-nesting areas are being monitored for oil spill effects on the northwest shoreline of Puerto Rico. This survey will help determine if there are any discrepancies in the normal amount of eggs which hatch, usually about 70-80 percent in a batch of 100 eggs per nest. The peak season for leatherback sea turtles in the Piñones area is April through May, and for hawksbill turtles from early summer into August. Any significant reduction in the hatches may cause the commonwealth to seek compensation for this loss in resources. When possible and necessary, USFWS has the authority to move turtle nests endangered by oily sand and relocate them to safe beaches.

The humpback whales that travel through Mona Passage off the site of the sunken barge, the manatees who live in caves off Shaks Beach on the western coast, and the thousands of inhabitants of the mangrove forests and lagoon leading into San Juan Bay were fortunate to largely escape the oil spill's swathe of destruction. Other organisms struggled for survival and, more often than not, lost.

Tony Mignucci considered the comparison value of one lost sea urchin to the value of a whale. He smiled and replied, "Maybe a hundred sea urchins make up one whale. It doesn't matter. We tried to save them all."[17]

SHORELINE CLEANUP
PART I: BEACHFRONT, SEA BOTTOM AND ROCKY COASTLINE

NO EASY OIL SPILL

The *Berman* onshore cleanup proved the old saying, "There is no easy oil spill." That was especially true for the part of the response focused on removing oil without harming the *natural resources*, which include all living resources, their habitats, and the land, sea and air around us.

Berman spill response applied to all natural resources, whether privately owned or managed by federal and commonwealth agencies. According to LCDR Brad Benggio, NOAA's Scientific Support Coordinator to the FOSC:

> *"The workers focused attention on the 'usual' subjects of a response: shoreline consisting of mainly sand or mixed sand and gravel beaches, rocky points, some wetlands, mangroves and small inlets; bird species of diving, wading and shorebirds, raptors and waterfowl with some rare and endangered species; fish and shellfish; the endangered West Indian manatee; and three types of endangered sea turtles and their hatchlings in sensitive turtle-nesting areas."[1]*

There were more unusual resources, those affecting people. As the FOSC said, "People are part of the environment, too."[2] The risks to socioeconomic resources were high, given that the spill occurred in a high use tourist area at Christmas time. These resources included: major tourism beaches, hotels, restaurants, casinos, shops, recreation activities, fishing, docking and marina facilities, and areas relating to the cruise ship industry.

Unique to *Berman* were the archaeological and historical resources, the forts and ruins along the oiled shoreline.[3] These *heritage resources* demanded much time and special attention.

FIRST AID

No two oil spills are alike. Berman was no exception to this rule. The job facing responders was tough enough without complications. Unfortunately, the shoreline cleanup was hampered by unique conditions which made the response more difficult than usual.

One early problem: the same stretches of shoreline were worked over many times as the barge continued to leak oil for nine days before she was finally removed. *Gross* recovery operations were conducted initially to remove floating oil in some area and beached oil from beaches along various sections of shoreline. The FOSC could not wait until the barge was scuttled to begin shore cleanup. No. 6 fuel oil is highly toxic to bottom and subtidal creatures, and to shallow water and reef fish. In the early days of the spill, there were large kills of fish, mollusk and echinoderms. The longer oil remained in the intertidal and subtidal regions of the beach and shore (areas near the shoreline), greater numbers of marine wildlife died. Rapid response was called for, even though this meant that crews faced the unpleasant task of partially reworking a beach already subjected to gross recovery efforts. Intensive shoreline cleaning of the most heavily impacted areas did not begin until the barge was removed on January 15, when the threat of *re-oiling* was essentially removed.

Responders labeled this early effort "First Aid work," cleanup performed in the initial hours and days with available equipment and personnel while additional forces are mobilized, usually concluded by Day 2 of a response.[4] On-the-spot-response equipment was limited initially to booming, Tyvek protective suits, and other cleanup gear. What equipment existed on hand was placed in critical positions across the San Antonio Channel to protect San Juan Harbor. Even after the majority of equipment arrived, in the early days some responders felt that they were "playing the response by ear."[5] The cleanup did not take shape as an orchestrated effort until the full incident command structure was in place.

BAD BEHAVIOR

The characteristics of the spilled No. 6 fuel oil hampered cleanup and contributed to recontamination in the early days. This substance floats, sinks and then resurfaces. Because of the wave action near shore, the oil mixed throughout the water column, picked up sand, became heavier, and sank. It took as little as 2 percent sand by weight to cause the oil to drop to the bottom. Oil on the bottom sticks together to form large patches up to 20 centimeters (8 inches) in thickness. Once there, the oil refused to stay on the ocean floor. The same waves that caused the oil to submerge, or higher air temperatures in the shallower lagoons, stirred up the petroleum and refloated it on the water in the form of slicks and tarballs.

The small *globules* on the ocean surface were described as resembling "chocolate-covered malt balls" by one chemist. CDR Stanton calls this heavy oil *"Berman* Balls." These tarballs do not disperse like other kinds of petroleum. Those that do not accumulate sand and sink again to the ocean bottom float up onto already worked beaches.

Tarmats blanketed huge expanses of seagrass on the lagoons near the Hilton and Radisson hotels. Until these mats and tarballs were recovered from the ocean floor, or wiped off the seagrass, the oil continued to resuspend in the water column and flood the shoreline. Recovery of subsurface oil became a priority.[6]

No. 6 fuel oil surfaces as a solid tar-like mass that hardens as it weathers. The oil spread over the hard substrates, rocky shoreline and manmade riprap rock embankments protecting hotels and beachfront area. This viscous liquid reached heights up to 25 feet from waterline in some cases, where it covered historical artifacts dating back to the 15th, 16th and 17th centuries. Removal of the encrusted oil required specialized cleaning agents, dispersant, and the oldest cleanser of all, "elbow grease" applied by manpower.[7]

Unfortunately, economic forces are making this type of heavy oil more and more common in the industry. Utilities and many industrial companies use the family of residual oils referred to as LAPIO (Low API Oil, a very heavy residual fuel oil), in greater frequency because these oils provide "the most BTUs for the buck."[8] *Berman* serves as a case study for future anticipated LAPIO spills.

SAIT

Capt. Ross squared off against the oil recovery with dispatch. He divided the shoreline into 16 zones. Two additional zones were added on the western coast of the island after the second release.[9] The FOSC delegated the details to competent people. He organized a *Shoreline Assessment/Inspection Team* (SAIT) to protect and clean up natural resources in the affected areas. Each of the involved natural resource trustee agencies were members of the team: Puerto Rico Environmental Quality Board (EQB), Puerto Rico Department of Natural Resources and Environment, USFWS, National Park Service (NPS), NOAA and the USCG.

TRACKING

NOAA's Brad Benggio advised the FOSC on "all scientific and environmental issues" affecting the oil spill cleanup, and represented NOAA's trusteeship to assist in the assessment of damages caused by the spill. When Benggio arrived at the scene, he was asked, "What do you think the oil will do?" He answered, "I don't know. I've never been in the same spill twice."[10]

To be fair, the question was not as naive as it sounds. One of the primary problems facing the team of NOAA experts during their two months on the scene was *tracking* the oil, trying to predict where oil would end up so workers could be placed properly to accomplish the most effective cleanup.

Tracking oil is difficult because of heavy fuel oil's common sinking and emulsion traits. One observer described the oil as "running all over the place." Spilled oil could travel inches or miles underwater, driven by one set of currents and then another, and show up in a totally unexpected place. Unfortunately, a "beach is the best oil barrier."[11] When oil finally hit the beach, the workers could deal with it. But a big problem in the initial days was trying to place workers onshore and say, "This is where the oil will emerge."[12]

NOAA flew over the spill nightly. A USCG member described the flights:

> *"We'd fly daily, one time during the day and one time at night. We'd be able to see how much the beaches were impacted. In the case of the later oiling of Aguadilla area beaches, there were whales in that area, and we had to keep watch over them. NOAA used infrared cameras at night. These would produce a picture that showed oil on the water as 'bright stuff.' NOAA would report this added information to headquarters to help them track the fuel oil."*[13]

NOAA's team tracked the oil for six weeks using complicated computer models supplemented by local knowledge to provide the FOSC

Aerial of shoreline skimming. (Reprinted by permission of U.S. Coast Guard MSO, San Juan.)

with scientific support. NOAA used two methods to project the path of the oil. *Trajectory Modeling* involves state-of-the-art computer work. The other method is the tried and true "hands on" practice, using local knowledge. At their headquarters in Seattle, WA, NOAA runs On-Scene Spill Models (OSSM) on main frame computers. Using this data base, NOAA adds local knowledge gained from daily flights, including where the oil is, what it looks like, and how it appears to be moving. In *Berman,* the major direction of oil flow was west. But local eddies near Escambrón Point caused some oil to move east. Once the NOAA team relayed this information back to Seattle, the main frame kicked out a trajectory model telling the FOSC where he could best expect the oil to flow. Based on these predictions, the FOSC and his team could schedule workers to respond and set up advance protection for sensitive environmental areas.

According to Benggio, the sophisticated tracking proved to be surprisingly accurate.[14]

COREXIT 9580 - ASPHALT PREVENTION

NOAA used a computer program called the *Oil Fate Model* to design response. By entering the density, viscosity and volume of the oil, along with local information regarding wind, waves and water temperature, NOAA can predict how much oil will evaporate from the surface and how much will disperse into the water column.

Depending on how close the oil is to the shore, the type of water, reefs, mangroves and other environmentally sensitive areas in the immediate vicinity, NOAA scientists project the best method of responding to the oil. Scientists use chemical treatments for oil in three broad categories. *Dispersants* break free floating oil into small particles that spread in three-dimensional fashion into large droplets. Unlike the previously suspended oil, these particles do not join together again to form tarballs. The weather, waves, wind and sun dissipate the dispersed oil over larger areas until microbes in the water can effectively ingest it. *Chemical cleaning/releasing agents* act much like soap, lifting the oil from surfaces and causing it to refloat. *Surface pretreatment,* the least proven method, treats an area in advance of the oil hitting it. In *Berman,* the FOSC chose to use chemical cleaning/releasing agents for surface cleaning.[15] An old pollution control response maxim is: "The solution to pollution is dilution." Dispersants do just that.[16]

In *Berman,* NOAA chose the chemical compound *COREXIT 9580* after a series of careful tests. COREXIT 9580 is designed to "lift, float, and recover" oil. First, affected areas were cleaned with *sorbent snare,* plastic-like material that resembles cheerleader pompons, to which oil adheres. COREXIT 9580 was applied and allowed to soak into the black stain that remained, and then workers blasted the rocks or shore material with

Shore cleaning. (Reprinted by permission of U.S. Coast Guard MSO, San Juan.)

high-pressure water at a controlled temperature. This washing removed oil from the shoreline surface to the sea, where sorbent boom caught the released oil, or where the oil dispersed in small amounts in the sea.

Marketed by Exxon, COREXIT 9580 is supposed to be biodegradable, and is one of the EPA-scheduled, acceptable products for spill cleanup. NOAA selected Vance Vicente, an oceanographer and longtime Puerto Rico resident, as chief scientist to help determine how to protect biological resources and to assess damage to the natural resources affected by the spill. Vicente was asked to determine how toxic COREXIT 9580 was to marine creatures in the shoreline area to be cleaned.

The testing procedure was crude but expedient, as circumstances warranted. Exposed animals were collected from tide pools near rocky shores where small amounts of COREXIT 9580 had been applied. Then the animals were placed in several sea water aquariums on Vicente's desk in the Command Center. If the animals survived, Vicente deemed COREXIT 9580 an appropriate agent for use in breaking up oil.[17]

The primary method of shore cleaning was by hand, using a shovel, rake and screen sifter. In some locales, machinery, tillers and small bulldozers did the work of many men. Conveyor-driven separators were used in other areas. Dispersant use was reserved for two types of environments:

- Areas where the FOSC's team determined that shoreline was of high public use, and quick and more thorough cleaning was desired.

- *Low energy* areas, where natural and mechanical cleanup could not prevent the formation of *asphalt*, or hardened tarmats or pavement. [18]

HARDLY A VACATION

Manpower, muscle and strenuous labor, not chemicals, computer models or high-tech gear, cleaned most of the shoreline. At the height of the cleanup there were reportedly 2,000 workers (other sources say 1,000) in all phases of the operation, under a core team of supervisors.[19] Most of the workers were local residents of the commonwealth. The FOSC's Spill Management Team supervised the workers, and the EQB monitored their employment. Brad Benggio assisted the FOSC through his Shoreline Assessment/Inspection Teams (SAIT).

SAIT members pulled tough duty. Mayra Garcia, a representative of DNR from San Juan, worked on a SAIT team and described the daily drill:

> *MAYRA GARCIA: I was one of the lucky ones who got to walk the beaches daily and conduct shoreline evaluation. We were suited up in coveralls, safety boots and over-booties for protection. We wore all this gear for five to six hours of beach and shoreline walking, which we did each day, under a blazing sun and 80 to 85 degree F. temperature. You can imagine what that was like.*
>
> *What we'd do is meet each night and review the cleanup progress. We'd make a plan about which areas each of us would cover for the next day. What we were looking for initially was how much oil was on site and what method was being used for the cleanup. If there was a sensitive area, we'd check to see if the workers were being careful of critical habitat. We needed to know what had to be done to complete the cleanup.*
>
> BARBARA E. ORNITZ: Did your job change at any time?
>
> *MG: Yes, as the cleanup progressed. At first we would simply walk the beaches and fill out a preprinted form itemizing what the area looked like and the stage of the cleanup. Later, it was my job and that of others on the team to sign off on whether the contractor had cleaned the area sufficiently for us to move the crews elsewhere. Every few feet I had to take my shovel and dig in the sand. We had guidelines about how clean the spot should be, like so many grains of oil the size of a dime or pinhead per so many feet of sand. I'd dig, examine the sand, see if our cleanup standard was met, and move on. If the area wasn't clean enough, I'd schedule a review for the next day or the day after. Each night the team would go over these reports and put the particular area on the schedule for another review or mark it off as clean.*

> *I walked the shoreline for miles daily from January 7 through the first of March, except for one week when I got the flu. That was hardly a vacation from what was at best a gruelling two and a half months.*[20]

Scientist Aileen Velazco-Dominguez, the DNR Scientific Support Coordinator and chief advisor to Genaro Torres, the head of Puerto Rico's EQB, organized DNR's response efforts. She remembers receiving notification of the spill early January 7:

> *"I grabbed my oldest clothes, knowing I was going to be throwing them away later. I put on my old sneakers and headed out the door. We worked long hours daily and covered hundreds of kilometers in a day walking the beaches and doing our assessment work. At night we met and decided what we would do for the next day, reviewed the efficiency of the cleanup, and headed home about 11 p.m. At the end of three months, I was exhausted physically and emotionally. Even so, you can't turn your head off during a spill. I saved that for later."*[21]

For years, Aileen Velazco has conducted studies for her department on coastal erosion and marine geology. Her knowledge of the affected areas was invaluable, and her recall close to photographic. One day while surveying the Piñones area, she discovered to her horror a new pocket of oil not previously reported. Piñones is the site of some very sensitive mangrove forests. Aileen notified NOAA at the command center. So accurately did she describe the exact location of the spill and the status of the oil that by the time she returned, the Coast Guard already had a crew on the site and a boom in place to save the lagoon. [22]

OCCUPATIONAL HAZARDS

La Perla is an infamous "city within a city" in San Juan, a squatter's development on beachfront owned by the park service near the heart of Old San Juan and Fort San Cristobal. Years ago, the city's poor and homeless moved in and took up residence in a sprawling maze of ramshackle houses and beach debris. La Perla became a haven for drug dealers and other notorious characters, a place where decent citizens feared to tread even in broad daylight. The National Park Service (NPS) began an aggressive campaign to cleanup La Perla. At the time of the *Berman* spill, La Perla was still an unsavory spot to visit.

At one point during the response, workers were engaged in cleaning up the beach and large rocks at the base of La Perla's jumble. Responders had gathered 12 or 13 bags of tarred, oily debris, a dangerous and noxious-smelling substance. These bags were piled under one of the more substantial houses in the neighborhood, to be collected at the end of the cleanup. Suddenly, workers found themselves confronted by several large, unpleasant-looking thugs brandishing weapons who pointed straight at the beach crew. "Get out now," they commanded. The workers obeyed without dispute and left the bags of toxic debris where they lay.

The crew had unknowingly picked the wrong spot to store their debris. The house they had chosen belonged to a shady underworld character, who clearly objected to any strangers snooping around his home. Not long after, fumes from the oiled material began bothering residents. Weeks later, the citizens were sufficiently concerned about the mess at their doorstep to allow response workers back. Freddie Aledo, a NPS ranger, explained to residents what the crews were doing and why they needed to be on the beach area below La Perla. Thereafter, each crew was accompanied by an armed ranger. Cleanup of this area posed an unforeseen occupational hazard not normally encountered in the average spill response.

After much prodding, Aledo reluctantly provided a brief tour of La Perla. He was rightfully uncomfortable about taking a visitor into this inhospitable squatter's city, and kept his hand close to his holster. He contacted NPS employees by radio when we walked down to La Perla, and let them know when we had returned safely. Aledo pointed out the dealer's house from the cleanup episode, although he need not have bothered. In a realm of cramped lean-to huts, broken windows, open pipes running into the sea and general tenement conditions, the gangster's neatly stuccoed, fully glassed pastel green residence was plainly evident.

The beach and rocky shore in front of La Perla were "clean enough for government work." Some dark streaks remained in and under the rocks. Aledo dug a deep hole in the beach to reveal a small amount of oil seeping through the sand. Yet, all things considered, the beach looked reasonably clean.[23]

BLACK SAND

The image that confronted Nelson Pereira of the USCG was far from clean when he first arrived on Piñones Beach. Pereira was a leading Petty Officer in the Marine Environmental Response Branch at the Marine Safety Office (MSO) in San Juan during the spill. His job included tracking and monitoring spills reported to MSO San Juan. Since 1991 he has been responsible for overseeing the work of private contractors hired by RPs to clean up oil discharges. He served in that capacity with others in *Berman*, dealing with NRC, Crowley, Endochem, Caribe Hydroblast and Oil Mop from the U.S.

Pereira was assigned various regions during the spill, and covered Piñones, on the east side of the spill impact; Culebra, an island off the northeast coast of Puerto Rico (part of the mystery spills); and Aguadilla, the site of oiling from the major release that occurred after the sinking of the barge. Said Pereira:

> *"When I saw Piñones, all I could see was black sand. Higher off the sand there's a sensitive turtle-nesting area. It was our job to plan the attack against this oil without impacting this sensitive place. There was so much oil on the beach, and we had to do everything by hand, that at one point we used over 200 people. The contractors brought in supplies from the states, hired local people, gave the personnel a basic four-hour training course about how to protect themselves from heat and oil exposure, and sent them out with picks, shovels, plastic bags and sifters."*

Was the beach clean after all this effort? "Compared to what it was like, yes," said Pereira. "There are such small tarballs remaining, we'd do more damage than good trying to clean them up."[24]

HOW CLEAN IS CLEAN?

Brad Benggio considered the question of how close spill responders can come to achieving perfection:

> *"We knew we couldn't remove all the oil. In some places if we scrubbed too hard, we would do more damage to the environment than if we let nature take its course. This left us with the job of developing realistic guidelines about how far we needed to go to consider a particular area clean."*[25]

Capt. Ross confirmed this approach. His concerns were safety-oriented:

> *"When you ask, 'How clean is clean,' your answer must include aesthetic, biological and political considerations. Does it make good sense to put people at risk to remove oil in a non-biologically sensitive area where nature can take over in time? For example, the rocky shoreline to the west of the grounding site is a high energy area. Waves break over the rocks and weathering will clean the oil over time. In other areas, there might be a higher standard that has nothing to do with environmental issues, but is necessary to prevent economic impact."*[26]

The FOSC and SAIT applied a common sense view to develop a set of written guidelines for contractors describing how clean each area must be before responders moved on to the next location. The instructions were very specific. For example, on certain beaches "clean" meant that the tarballs left on the beach could be no larger than the size of a thumbnail and there could be no more than 10 per square centimeter.[27] The goal was results rather than perfection. A typical guideline confirms the reasonable approach taken by the responders:

> *"Beachrock – Areas of high recreational use: Heavily oiled natural beachrock in areas of high recreational use will be cleaned using shoreline cleaning agents and high pressure/hot waterk flushing. Only one treatment with the cleaner and pressure washing system will be conducted ... to minimize the potential for damage to the beachrock. Even after cleaning, there will be black oil on the undersides of the rock, in crevices, and some patches on the surface, and extensive oil stain. The objective is not to remove all of the oil, but to remove the gross oil so that natural removal will be enhanced."[28]*

When a contractor signalled completion, the beach supervisor inspected the area. Then SAIT members walked the beach to determine if standards had been met. The FOSC reviewed the SAIT recommendation and gave his final approval. Once an area was considered clean, crews moved on. The FOSC involved commonwealth officials whenever possible in final inspections of the island's beaches and shores.

Three months to the day of the spill, the last of the Escambrón area beaches were deemed clean. Susan Soltero, a well-known Puerto Rican broadcaster for *TeleOnce* (Channel 11 News), refused to take the USCG's word that these previously black-oiled beaches were once again clean white sand. Dressed in her full business suit, Soltero took her TV crew down to the beach for a look around. Delighted with what she saw, Soltero kicked off her shoes and waded into the waters of the Hilton Lagoon. She showed the bottoms of her bare feet for the TV camera, smiled with joy, and announced, "It's really clean."[29]

ASPHALTING THE OCEAN FLOOR

As the No. 6 fuel oil collected sand and debris, it became so thick that it sank to the ocean floor and lay there like strips of asphalt on the sandy bottom. There were several principal areas of submerged oil: in the lagoons near the Caribe Hilton and the Radisson Normandie, in the area between Escambrón Beach and its fringing reef, in the shallows north of Dos Hermanos Bridge at the entrance to Condado Lagoon, and in the reef shallows at the west end of Shaks Beach near Aguadilla. Fields of seagrass covered the Condado Lagoon entrance by the Dos Hermanos Bridge. Cleaning this seagrass required individual attention to each blade. In the Aguadilla area of sensitive coral reefs, seagrass beds and manatee caves and holes, the oil lay on the bottom of the ocean floor in thick patches.[30]

Recovery of submerged fuel oil is no easy matter. The operation

Diver collecting *Berman* tarmats. (Provided by Tom Eason of Eason Diving & Marine Cont., Inc. in Charleston, South Carolina.)

required many people, much equipment and money. The total cost for submerged *Berman* oil recovery is unclear. One estimate was $8,200,000 to net about *145,000* gallons of oil – a recovery cost of slightly more than $50 per gallon.[31] This figure was revised to $9,560,000 in the FOSC's August 1995 report, increasing the cost to $66 a gallon. Arguably, while this was not a cheap or efficient operation, the longer term consequences of leaving the beaches subjected to re-oiling from submerged oil was not an acceptable outcome. The submerged oil had to be recovered. Again, Capt. Ross faced a unique problem. The science of submerged oil recovery is in its infancy, with little knowledge industry-wide. The FOSC appointed a task force of experts to deal with the problem.

Their approach was to use vacuum transfer units and other mechanical devices wherever possible to suck up the oil. When machinery would not do the trick, the team employed divers to remove the oil manually. They dived into the murk and vacuumed up oil until it clogged their hoses. The remainder was collected by hand. Divers placed the tar-like mats by hand into onion bags, which were then hauled to the surface.

Acting as dive master under contract to the USCG, Tom Eason of Eason Diving and Marine Contractors, Inc., Charleston, SC, supervised the submerged oil recovery diver operations. Incorporation of the contaminated diving and decontamination procedures developed by the company on previous projects resulted in the successful completion of oil recovery efforts in 60 days with no Eason Diving injuries and minimal loss of equipment resulting from contamination.

Sand cleaning in the lagoons involved two dredges, one from Puerto Rico and another imported from Wyoming. The dredges pulled in tremendous amounts of sand and oiled water. To clean the sandy, oiled water, the FOSC resorted to an ingenious system. Responders used a swimming pool left over from the Pan Am games and three smaller pools to *decant* oil from the water, remove the sand, and return filtered sea water back into the ocean.

Diving and vacuum removal took place January 22-25, and lagoon dredging and *decanting* began January 27. The whole process consumed some 96 days.[32]

STAGGERING COSTS

Three companies provided over 50 divers and supervisory personnel to recover submerged oil and sunken oiled sorbent snares, and to clean seagrass beds. Except for the shallow depths, all other conditions made for an almost impossible situation. Because of the fuel oil's high toxicity, divers typically wore a rubber wet suit, covered by a protective, specially designed Tyvek suit, a full head helmet that looked like a large hard hat bubble, and steel shanked boots for better footing. The head helmet alone weighed 34 lbs., and divers carried 60 lbs. of weight. Usually divers had an "umbilical cord" attached to their suits for their air supply. Most helmets were rigged with communication gear, including underwater video equipment in some instances.

Added to the weight of this rig was the cleanup gear carried by each diver. Usually this consisted of small suction pumps with hoses attached to compressors at the surface, or heavy onion bags. In the case of seagrass, divers used sorbent snares to free individual blades.

Handling hazardous waste required extreme concentration. Divers had to bring all material to the surface or connect their bags of oiled material to a grappling hook connected to a boat

Berman diver from submerged oil recovery operations. (Provided by Tom Eason of Eason Diving & Marine Cont., Inc. in Charleston, South Carolina.)

at the surface. In some cases the oil was a foot deep and easily stirred up, leaving the diver surrounded by an underwater cloud of black muck. Divers also had another health consideration. They had to avoid human waste, hypodermic needles and broken bottles in each of the places they were diving.

Once on the surface, decontamination of divers became a priority. They removed their suits and then washed down thoroughly. The divers used a "step out" procedure, much like that employed in *nuclear* diving. Responders handled the equipment with equal care, given the staggering costs of replacement. One helmet costs about $3,700, one Viking drysuit about $2,000, and the underwater video system, camera and umbilical cord connecting the diver to surface air, monitoring and communications systems cost about $12,000.[33]

SHORELINE CLEANUP
PART II: SAND, SLUDGE
AND SPECIAL PROBLEMS

Berman posed a unique situation in terms of recovering oil from the bottom of the two impacted shallow lagoons, disposing and treating debris left over from the cleanup, and the specialized care of historical artifacts in the path of the oil.

PAN AM POOL
Separating out water from oil and dredged sand and returning clean water back to the ocean taxed the ingenuity of the responders. The only effective way to remove oil from the lagoons behind the Hilton and Radisson hotels was to dredge up the oil and siphon the sand/oil mix through a hose running out of the lagoons to a separating process that sorted sand, water and oil. While dredging is an uncomplicated process conducted all across the U.S. by the Army Corps of Engineers, once again *Berman* posed unique problems.

Dredges work by tilling up the sand, cutting a one-foot-deep swath and pulling up sand, oil and water through high volume pumping stations. This mixture was pumped through a hose into treatment pools. To prevent

Lagoon cleaning. (Reprinted by permission of U.S. Coast Guard MSO, San Juan.)

Decanting pools. (Reprinted by permission of U.S. Coast Guard MSO, San Juan.)

rocks from damaging the cutters or pumps, rock guards were installed on the dredges and divers scouted the path in front of the machinery to remove rocks. Sand caused extreme wear and tear on movable parts and hoses. Fittings and pump impellers required constant replacement.

In spite of these setbacks, tremendous amounts of contaminated material moved through this dredging system from February 1, 1994 to April 13, 1994. In just 16 days during this period, workers dredged about 2,053 cubic yards of oiled sand. The recaptured oil went from the site to Barge *250-1* and then to Sun Oil Refinery in Yabucoa. Workers hauled the oiled debris and sludge which could not be cleaned further to the Browning Ferris Industry (BFI) landfill in Ponce, on the southern coast, along with 20,000 cubic yards of semi-contaminated sand.

CDR Stanton came up with a creative way to employ local resources to return near-clean, treated water to the sea. He "requisitioned" one warm-up swimming pool in an athletic complex by the ocean, built in San Juan when the commonwealth hosted the Pan Am Games in the late 1960s, and placed one 50,000-gallon stilling pool and two 12,000-gallon stilling pools in the same complex. This set-up was conveniently located 100 yards from dredging operations in one of two oil-filled lagoons in the vicinity of the Caribe Hilton and Radisson Normandie hotels. Stanton seized upon the idea of using these pools and a series of washes to separate and then *decant* the sand, oil and water.

The USCG invented an unusual oil/water/sand separator, applying the basic principle of gravity. The small warm-up pool, about 140 feet long by 70 feet wide, served as a separating basin for the water/sand/oil mix dredged from the lagoons. The highly agitated mix was discharged from the dredge pipes into the pool. The sand sank to the bottom. Newly freed oil floated to the top, where it was recovered with a weir skimmer and fed through hose to a vacuum truck. The sand was removed with a backhoe and piled alongside the warm-up pool, allowing the water to further separate and drain back into the pool. The wet sand was stored in a *roll-off box*, a canister much like a trash dumpster, then loaded into trucks and taken away to the Ponce BFI landfill.

The water (still containing oil) was pumped through the other pools

OIL/WATER/SAND SEPARATOR AND DECANTING SYSTEM

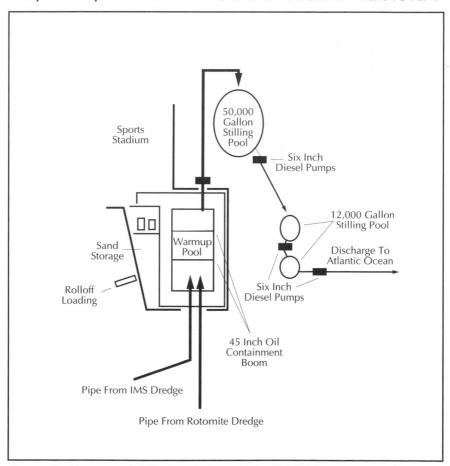

(Source: "The Response to the T/B *Morris J. Berman* Major Oil Spill," San Juan, Puerto Rico, USCG. Reprinted by permission of U.S. Coast Guard MSO, San Juan.)

where the oil was removed from the water. Sorbent material in the three portable pools collected almost all remaining oil. A vacuum hose from a truck skimmed any remaining oil still floating on the surface of the stilling pools. As recovered water moved from pond to pond, the oily residue decreased. The USCG's ingenious cleaning system worked so successfully that water from the last pool could be discharged directly into the ocean with the approval of the commonwealth, EPA and all agencies involved. Almost three million gallons of "clean" water was returned to the ocean by this process.[1]

PAC MAN WITH A MUFFLER

FOSC Ross and his staff next faced the question of what to do with thousands of cubic yards of dredged sand. Waste disposal experts from BFI pre-positioned roll-on/roll-off boxes at various cleanup sites on the north shore of Puerto Rico to receive solid waste and oil-stained sand from beach cleaning and dredging operations. Crews worked around the clock to transport these contaminated substances to the BFI landfill in Ponce, where sand was segregated by site of origin into five different piles so that each type of sand was cleaned separately.

- Dredged sand from the Hilton/Radisson lagoons.
- Piñones Beach sand.
- Hilton Beach sand.
- Escambrón/Reserve Officer's Beach sand.
- Aguadilla sand.

Thousands of cubic yards of sand arrived at the Ponce landfill, where BFI used the natural process of *bioremediation* to clean oil from the 4,000 tons of sand. BFI built a clay pad to hold the sand and keep oil from percolating back into its natural storage tank, the earth. Spreading the sand on top of the clay pad 14 to 16 inches deep, they "fertilized" the sand by mixing it with nitrogen, phosphates and other nutrients — similar to mixing compost in your garden.

BFI then added *oil-hungry* microbes to the mixture. These microbes live naturally in the ocean, sand and soil. They break up hydrocarbons by feeding on elements in the chain of hydrogen and carbon that release gas, causing the oil to essentially disappear.

Rick Good, the industrial services manager of BFI's South Atlantic region, calls these microbes "Pac Men with a muffler" (referring to the popular video game in the '80s which featured a yellow caricature gobbling items for points). As long as the sand/oil mix is kept moist and turned to allow air into the substance, the microbes do their work. BFI only needed eight drums of microbes in the mix to bioremediate 4,000 tons of sand to acceptable levels in 45 days. The remaining oil particles were virtually invisible.

In fact, these Pac Men thrive on beaches. If the FOSC could have turned, fertilized and aerated the sand on San Juan's beaches, there would have been no need to truck the sand to Ponce. The commonwealth was unwilling to wait any amount of time or to have the USCG turn prime tourist beaches into "sand farms" for many months. This left the FOSC in the position of cleaning sand ("sand washing") on the spot, or importing sand from other high-use tourist beaches.

The cost to bioremediate oiled sand is between $60 and $120 per cubic yard. In *Berman*, the total spent on recovered oil transport, storage and disposal was $5,172,000. Sand washing alone cost $781,000.[2]

This was the first such bioremediation project undertaken by Puerto Rico. At the time of the *Berman* spill, there were no rules or regulations authorizing bioremediation in the commonwealth. Genaro Torres of the EQB hopes the Puerto Rico legislature will introduce necessary laws to allow future use of these mighty microbes.

The commonwealth owns the tons of cleansed sand sitting at the landfill. Since Puerto Rico is a "sand poor" island, according to one source, they intend to sell the sand to the highest bidder or use it for public works in construction rather than return the treated sand to the beach.[3]

SAND WARS

Not all sand operations ran so smoothly. The FOSC hit a major snag when he tried to replace the Hilton Hotel's sand. Responders removed about 1,000 cubic feet of sand from the Hilton beach. Because the spill happened in high season, the hotel was anxious to reopen as soon as possible and not wait for their beach to be cleaned through the normal course of sifting or bioremediation. The FOSC applied to DNR for a permit to replace the Hilton sand with material of similar color and grain size from the Piñones area. DNR approved a permit to remove 5,000 metric tons from the south side of the Piñones beach — not surprising since the commonwealth owns the Hilton.

Trucks set off for Piñones. Two loads of sand left the beaches before citizens there erupted in outrage. There was no way the Piñones residents were giving up their sand so the "rich tourists" could lounge on "clean" beaches. They formed a committee, organized themselves into groups and barricaded their beach.

This was a rare instance when the FOSC's people failed to handle communications with care. The Piñones residents did not understand that the sand being taken from their neighborhood was the sand piled on the side of the road opposite the beach — *not* their beloved beach sand. Had the FOSC informed the public of their plan, the Puerto Rico "Sand Wars" might never have happened.

The citizens eventually triumphed, and Piñones sand remained on their beach. The crews returned to the Hilton and sifted the sand from its own beach and lagoon, using 1/8th-inch-diameter sifters and a sand-washing machine. In this manner, the Hilton dredged sand from its lagoon and the adjacent beach stayed home.[4]

USING MOTHER NATURE TO TACKLE THE WASTE

The FOSC still had to deal with the sludge and other oily debris left over from the cleanup. While oil is not considered a "toxic" waste, under U.S. law it is a "contaminated substance" and must be handled sensitively. If oiled waste material was a "hazardous" substance, Puerto Rico would have been required to put all contaminated material into Dept. of Transportation-approved drums and ship the waste to authorized landfills stateside. The commonwealth has no approved hazardous waste dumps on its island, but it does have two approved disposal sites for controlled substance waste. One of these is BFI's facility in Ponce. BFI can handle up to 26,000 cubic yards of controlled substance in a dedicated *cell* on their site. *Berman* debris consumed 20,000 cubic yards of this cell.[5]

The *Berman* waste consisted of every item imaginable: polyurethene plastic "pompon" snares, absorbent booms, sludge, Tyvek suits (a protective plastic/paper fabric), gloves, rubber booties, personnel protective gear, mops, gross amounts of contaminated sand, oil chunks, and wood. The Tyvek and other protective gear used in *Berman* had to be thrown away after one use. BFI's Rick Good explained how they contained these waste products:

> *"You can't use plastic to hold oil. Hydrocarbons break it down. What you can use is clay, the natural material in which oil is normally found. Oil comes out from the ground. When we store it back into the ground as waste, we line the pit with clay.*
>
> *"Clay has a very high density. It's almost impermeable to oil. One foot of clay has the equivalent holding power of 100 feet of cement. It takes one drop of oil the same length of time to travel through one foot of clay as it does to pass through 100 feet of cement. So we used this natural material and lined the cell with 3 feet of clay. The other advantage to this substance is that it has great plasticity. If there's an earthquake, the clay will adapt and not crack."[6]*

BFI transported material from the spill scene to the Ponce landfill in two ways. The *Barge 250-1* traveled to Ponce, where workers mucked out the liquid sludge in the bottom of its tanks and trucked the sludge to the landfill. Once there, the liquid was loaded into a concrete vat lined with steel and mixed with kiln dust. Liquid seeps more easily through the earth, but solids stay in place. The kiln dust, usually pulverized cement, turns the oil into a solid substance better suited to cell storage. Workmen mix the dust and oil, then remove the resulting compound and place it into the cell. The second method involved BFI trucks transporting the solid waste, which was not subjected to the kiln dust treatment. Some 60 BFI employees worked 60-hour weeks for months on end handling spill refuse.

The *Berman* waste fills a cell the size of a football field, completely enclosed by a high chain fence. There is no waste visible because BFI has stored other substances atop the *Berman* waste since the spill and buried that waste under dirt.[7]

HERITAGE RESOURCES AT RISK

By the third day of the spill, the FOSC's staff realized that oil threatened historical structures. San Juan is a city of major historical importance. Prior to the 15th century, Spain occupied the island of Puerto Rico. On four separate occasions between 1595 and 1797, various European nations tried to seize control of this important harbor and fertile land. The Spanish built a system of defense fortifications on the eastern sector of San Juan Islet. Only portions of the 3.5 miles of historic fortress walls surrounding the 475-year-old city of Old San Juan remain.[8] Certain of these structures sat right in the heart of the oil spill and were seriously threatened:

- The Tajamar Ruins, an historic rubble wall on the Reserve Officer's Beach.
- Fort San Geronimo near the Hilton, as well as another historic stone rubble wall on the south side of the Hilton.
- An historic wall section on the south face of the west end of Dos Hermanos Bridge.
- Various pre-Columbian artifact sites (whose locations are kept confidential).
- Lighthouse ruins near the Aguadilla area.

Heritage resources are defined as "any known or potential historical or archeological structure, area or object within any cultural setting."[9] Berman responders used the term "heritage resources" because it is an internationally accepted term for World Heritage sites, and also out of sensitivity to political and ethnic concerns.

Natural resources may be replaceable, but there is little question that

16th and 17th-century historical structures are irreplaceable. Said one responder, "You can reproduce the Mona Lisa, but a copy is still only a copy and not the real thing."[10]

Like rare and endangered species, the U.S. treats heritage resources with great care. The National Historic Preservation Act (Title 16, USC. Section 470, 36 CFR Part 800) safeguards those structures either on the National Historic Register or those eligible for placement on that register. Section 106 of the act provides for elaborate steps and consultation between all trustees involved before any action can be taken with respect to an historic or archeological structure. The section's intent is to gain trustee agreement about how to avoid, reduce or mitigate any adverse effects to historic properties. Except for Fort San Geronimo, all the threatened structures were already designated as National Historic Sites. The San Juan National Historic Site (El Morro and San Cristóbal forts, El Cañuelo, and three miles of massive city walls surrounding Old San Juan) is also one of 12 NPS areas designated as a World Heritage site by the United Nations.[11]

In *Berman*, the trustees were the State Historic Preservation Office (SHPO), the Institute of Puerto Rican Culture (IPC), the U.S. Dept. Of the Interior (DOI) represented by NOAA, and the NPS. Agency representatives function as an advisory council, a semi-federal group. This group must meet, agree upon a strategy of protection and mitigation, reduce their deal to a formal document, and sign off.

To further complicate matters, the Area Contingency Plan failed to identify heritage resources on the area maps or provide methods for cleaning them. The only prior experience U.S. responders had with such heritage/historic sites was in *Exxon Valdez*. Yet the *Berman* and *Exxon Valdez* spills were entirely different in nature. In *Exxon Valdez* the climate was arctic, the affected structures were scattered far apart, and they were predominantly underground. There was little expertise available to *Berman* trustees about how to clean above-ground, natural sandstone, limestone or coral structures caked with No. 6 fuel oil in a tropical climate.

TEAMWORK

The FOSC responded quickly to this latest *Berman* dilemma. By Day 7 of the spill, he appointed a Heritage Resource Management Team (HRMT) led by USCG LCDR Audrey McKinley and assisted by NOAA's Brad Benggio. McKinley's team included members of the commonwealth's IPC, SHPO, the Maritime Bureau, Inc., the contracted spill management company, respected local archaeologists, and Dr. Agamemnon "Gus" Pantel, a local consultant and coordinator.

Gus Pantel is a longtime resident of Puerto Rico, a trained archeologist and anthropologist, and an international consultant. He has worked all over the world, and spent 20 years in Puerto Rico. His field is the sci-

BERMAN SPILL SITE MAP

(Source: "The Response to the T/B *Morris J. Berman* Major Oil Spill," San Juan, Puerto Rico, USCG. Reprinted by permission of U.S. Coast Guard MSO, San Juan.)

entific and administrative management of archaeological conservation of historic sites. McKinley described Pantel as "a man of many hats" who became "technical advisor, interagency liaison, team facilitator, coordinator and even field documenter."[12]

Dan Hamson, chief of environmental response, for the Planning and Assessment Dept., NPS, assisted local responders. Hamson wrote the first Memorandum of Agreement in the *Exxon Valdez* spill which enabled trustees to deal with the historical sites endangered by that spill. He contributed his expertise to smooth the way for various members of the HRMT to establish a simple set of guidelines for a Memorandum of Agreement. For example, before an historic structure could be cleaned, the team needed to agree in writing that a contractor could swab the structure with wet snare to remove oil. These and similar oil-cleaning methods were written and agreed upon in record time.[13]

The HRMT's job was to identify the impacted heritage resources, clean them, and then protect them from further spill damage.

CULTURAL BROKER

In crisis, timing and teamwork is everything.

Once the noxious odorous element (the aromatics) of No. 6 fuel oil evaporates, the substance develops into a resin that acts like a seal, waterproofing the structure underneath. This "band aid" effect tends to alter the homeostasis of five centuries. The substance crystallizes, creating a patina over the limestone or mortar works beneath. Gradually the walls and substrate material will crack and break down. The longer the oil stays on the structure, the worse the reaction is. Days become crucial.[14]

Such a crisis can create a situation ripe for confrontation between scientific and administrative approaches and two very distinct cultures. Scientists tend to take a conservative view and study an issue thoroughly before taking any action, particularly where there is no precedent to follow. Oftentimes a good scientist sees only the problem before him and acts in a myopic haze. In contrast, administrators look at the big picture from a budget perspective, and political expediency may result in accepting a less than perfect solution.

Like many of the *Berman* teams, the HRMT was a mixed bag of scientists and administrators, stateside federal representatives and commonwealth residents. Gus Pantel shared his insights about how he and others worked to solve the underlying dilemmas that typified different ways of thinking.

> *"The pace was incredibly hectic. We were on call 24 hours a day in the early days. Many of my professional colleagues had no appreciation for working long days and longer hours. At 4:30 p.m. or on a Saturday afternoon they would consider themselves finished with work, when we still had hours to go. Or we would need to make quick decisions about a certain matter, such as how to hot wash a wall or what chemical dispersant to use.*

> *"Scientists and military people like to run everything through the system, all the way up. The federal people distrusted local views. In one case, this meant bringing in an outside consultant and delaying the cleanup for days, just so we could watch him jump through the same hoops and come to the same conclusions as the locals had days earlier. This contrast in attitude was frustrating and time-consuming, not to mention costly."[15]*

The team managed to cut through their cultural biases and focus on solutions. Pantel describes his role as "cultural broker."

> *"In many cases you can't satisfy both sides fully and you can't achieve 'quantum compromise.' You have to take baby steps, find out what sacrifices and deviations people are willing to make from the perfect solution, and pick viable alternatives. You can't retrain people to look at the world differently during a crisis. You have to allow them to be flexible and use their multi-faceted skills. People are the*

Washing historic surface at Tajamar Ruin. (Reprinted by permission of U.S. Coast Guard MSO, San Juan.)

link. You work on forming partnerships. If you do the job right, no one feels coerced into anything. Every one walks away feeling like they came out ahead."[16]

PRACTICAL GUIDELINES

The HRMT team met several times daily, evaluated field sites, and issued decisions and written procedures for contractors. Procedures included what type of mechanical vehicles could be used at particular sites, how bulldozers could be operated in sensitive areas near historic structures, and what the contractor should do if any worker encountered an artifact. Contractors were required to fill out numerous forms indicating that they followed the carefully set methodology for working around artifacts. Team members reviewed the field work and approved or disapproved actions taken.

Procedures for protecting artifacts were practical and clever, including such measures as carpeting historic surfaces with sorbent material to prevent oiling, sandbagging ladders or equipment which might come in touch with these structures, constructing a simple foot bridge over historic walls for workers to use, and padding work boats with boom-like material to prevent chipping or breaking of walls. In every case, the team tried to find the least aggressive and most effective method for cleaning oil from the fragile walls. Careful testing led to an acceptable solution:

sorbent materials followed by hot water/pressure washing with operators holding a fan-tipped nozzle at a set working distance, and then
spraying water at a carefully selected temperature for a predesignated
time at the structure. Where that failed, the cleaning chemical compound
COREXIT 9580 was added. Sometimes cleanup required removing oil
that reached a height of 25 feet on the walls of Fort San Geronimo, leaving behind black ugly stains. Bands of oil one to four feet high coated the
surfaces of most affected historic structures.

Careful planning worked. Historical structures were restored to near
pre-spill condition without lasting damage. Precious artifacts were not
lost or removed. McKinley summed up her view of the heritage response:

> *"If you ask anyone involved with the* Morris J. Berman *oil spill
> response effort why it went so well, you are likely to get many
> answers. But in those answers you will probably find the recurring
> theme of community spirit and teamwork."*[17]

CHAPTER 7

VIEW FROM THE TOP:
MANAGING AN OIL SPILL

"How do you clean up an oil spill?" I asked the man in charge of the *Berman* response, Capt. Robert G. Ross of the USCG base in San Juan, Puerto Rico.

"How do you eat an elephant?" he replied. My mind conjured up all sorts of gargantuan responses that finally ended in an unknowing shrug.

Capt. Ross smiled and answered, "One bite at a time."

SO MANY PEOPLE

One of the first steps is organizing the troops. An oil spill response can be something like a party where too many people show up in too small a space in too short a time. The host of the party has to figure out what to do with all the people.

In the *Berman* spill, 15 federal and commonwealth agencies and about 50 separate contractors joined together to fight the spreading oil. Some arrived on the scene within hours, and others within the first weeks. Some stayed a short time, while the rest conducted their business through the next months until the end of April 1994, when the spill cleanup officially ended.

At the height of the spill there were approximately 1,000 federal, commonwealth, local and civilian responders combining their efforts to clean up the mess.[1]

FOLLOWING THE ROAD MAP

Once the troops are called up, they need to all march in the same direction and follow a prearranged plan if possible.

After the 1967 *Torrey Canyon* tanker spill near the coast of England, the U.S. began developing a national response plan. With the 1989 *Exxon Valdez* spill in Prince William Sound, public fury over the destruction of pristine Alaskan wilderness led to a substantial beefing up of little used laws and put teeth behind the plan.

The U.S. Environmental Protection Agency (EPA) and the USCG are two federal agencies that oversee oil spill responses, authorized by Section 311 of the 1971 Clean Water Act and the Oil Pollution Act of 1990 (OPA 90). Additional laws and international treaties affect the oil spill business, but these are the major pieces of U.S. legislation that currently apply.

83

These acts allow EPA or the USCG to take charge when an oil discharge occurs and establish supervision and coordination of response efforts, who is in charge, and how to best work together. EPA handles the *inland zone* and the "USCG manages the *coastal zone*. A series of plans on the national, regional and local levels organize the spill response business. The National Contingency Plan (NCP) is the umbrella document for activating a response to oil spills and hazardous substance releases. The regional and area plans are federally produced as well. A *region* is a multi-state grouping for administering agency programs at a level between the state/local and the national levels. The Area Contingency Plans (ACP) are the result of a collaborative effort between Federal, state and local government officials as well as special interest groups. The purpose of all plans is to coordinate a quick and effective spill response, with the NCP setting policy at the highest level, and decision-making proceeding at lesser levels of authority on down through the other plans. The Federal On-Scene Coordinator (FOSC) from either EPA or USCG is in charge.

When a Spill of National Significance (SONS) or a smaller spill occurs, the National Response System is activated, orchestrating the network between federal, state and local agencies, private sector agencies or individuals, and the RP.

According to the ACP, a SONS is categorized as:

> "*A spill which greatly exceeds the response capability at the local and regional levels and which, due to its size, location, and actual or potential for adverse impact on the environment is so complex, it requires extraordinary coordination of federal, state, local and private resources to contain and clean up.*"[2]

Berman was clearly a SONS spill.

At each level of this system, oversight committees run the show. On a national level there is a National Response Team (NRT) composed of 15 federal agency representatives. The next level down are Regional Response Teams (RRT) composed of federal and state representatives from each region. On the local level, an area committee acts as a planning body to help develop and maintain the Area Contingency Plan.[3]

In *Berman*, the players included the Caribbean Regional Response Team (CRRT) composed of federal and state agencies, EPA, USCG, DOI, and representatives from Puerto Rico and the U.S. Virgin Islands. The local agency charged with acting on behalf of the commonwealth is the Environmental Quality Board (EQB), with direct assistance and participation from Puerto Rico's DNR.

The RRT drafts the local spill response plan, called the Caribbean Area Contingency Plan, a long and detailed plan running hundreds of pages that:

- Identifies the people and agencies involved with a spill.
- Provides the names, addresses and telephone numbers of those to contact in the event of a spill.
- Specifies environmental areas throughout Puerto Rico and the Virgin Islands, with identification of key sensitivities in each of those areas.
- Highlights the equipment and other resources available in the Caribbean to fight a spill.
- Directs disposal of waste and hazardous materials produced in the cleanup.
- Provides logistical detail to support a response.
- Develops spill models for most probable, maximum most probable, and worst case discharges.[4]

OPA 90 preassigns the roles of all individuals who respond to a spill. The vessel owner or charterer is identified as the Responsible Party (RP), who is represented on the scene by a Qualified Individual (QI), the agent on the spot whose words and actions bind the RP. Usually the insurance underwriter for the vessel holder follows the cleanup closely and assists where necessary.

Each vessel owner in the business of transporting oil must have a vessel response plan. The RP's response plan designates one of the most important players, an Oil Spill Removal Organization (OSRO). The RP is charged with managing the response, deciding what to do in concert with the FOSC. The OSRO follows the orders of the Spill Management Team employed by the RP. The plan lists what the OSRO can do and how the OSRO will handle each of the cleanup functions: lightering, salvage, fighting fires, shoreline and offshore skimming, and payment to all those who assist[5] (See *Morris J. Berman* Response Management Organization flow charts, Appendix A).

THE NO-SHOW RP

Response under OPA 90 is not supposed to be top-heavy. Upper USCG or EPA management is limited in number. The RP should have a smoothly functioning and effective response management team, with an overall coordinator and an OSRO capable of delivering sufficient equipment and personnel to the site to skim and store a specified minimum quantity of oil each day. The OPA 90 model assumes the RP will show up and do its job from start to finish.

Responsible Parties come in all shapes and sizes. Some are well organized and financially sound, like the major oil companies. These RPs are usually "ready, willing and able" to provide effective management during a spill. A second type of RP includes single shipowners who want to be effective spill managers but lack the financial ability or expertise to manage an oil spill. The third kind of RP is a "mixed bag" of companies who have reasonable financial ability and some degree of readiness, but lack the will to respond. In *Berman*, the RP was a corporate entity which was one of numerous corporate fronts through which the notorious Berman/Frank family did business. This entity was unwilling or unable to assume an active role in response coordination.[6]

In the model situation the RP, not the FOSC, directs the cleanup contractor's actions. Management spheres are divided among these three. Typically the FOSC has no control over what the RP or his OSRO spends. The FOSC sets performance requirements (for example, "how clean is clean") that the RP should meet. The division of labor and responsibility calls for tight communication between the various parties who function in separate but often overlapping spheres. However, in the daily handling of crisis activities, sound strategies and clear communication can fall between the cracks despite everyone's best efforts at responsible management.

The party responsible for the *Berman* spill failed to show up. The companies and players involved caused the accident by their negligence, and then bowed out of the picture after the $10 million insurance ran out and the USCG took over. Pedro Rivera, the agent for the RP, responded immediately after *Berman* radioed for help. Thereafter, he kept a low profile through the initial response stages, testified at the USCG's casualty investigation hearing, and disappeared from the scene.

A scant month before the *Berman* spill, the Bunker Group's *BGI Trader* discharged some 10,000 gallons of No. 6 fuel oil into San Juan Harbor in Puerto Rico. Because the Bunker Group also owned *Berman* and because of problems with that incident, the FOSC knew this RP would be a problem and planned for the USCG to assume full control of the response from Day One. Neither the RP nor New England Marine Services provided the response management team identified in their Vessel Response Plans submitted to the USCG for approval prior to the incident, causing the insurer to hire Maritime Bureau, Inc. (MBI) to act as the spill management supervisor. When insurance ran out, Capt. Ross contracted for MBI to stay on and assist the USCG.

Since OPA 90 never envisioned the USCG being the primary spill responder, the USCG is not equipped with an organized management team. The Puerto Rican USCG MSO in San Juan simply did not have the manpower or training to take on the type of broad-based management required in *Berman*. Hiring MBI was only one of several measures taken to solve a severe staffing shortfall and service level response deficiencies facing the FOSC.

With so many players and untried pre-planning, many logistical details were carried out for the first time during the cleanup. For the sake of spill response effectiveness, much valuable time and energy was spent on communication, dividing up areas of responsibility, such as lightering, salvage, skimming, shoreline recovery and submerged oil recovery, and agreeing on procedures, rather than actually cleaning up the oil discharge and waste. Sometimes efforts ran parallel and were redundant instead of flowing neatly from the top command post on down with a unified point of view. For example, MBI and the USCG both managed shoreline cleanup. Capt. Ross describes the problem, which underlay the entire *Berman* response:

> *"This was a deliberate, though only partial, duplication because there was a significant lack of comfort in some Coast Guard circles with the concept of contracted spill response management working as part of the FOSC staff. The Coast Guard wanted to monitor the managers along with the cleanup contractors and wanted a second audit mechanism. The Coast Guard also wanted a backup management team in place in case the contracted management team did not work out. However, the backup team was not needed and ... the original unease on the idea of contracted spill response management was unwarranted."*[7]

As cleanup efforts progressed, these problems were solved. MBI assumed a larger role and the USCG monitoring staff decreased.

COMPETING INTERESTS

In a highly visible spill like *Berman*, the bottom line is different for each major participant. The public is outraged and focuses on the economic impact of such a spill and the health effects on visitors and residents. The RP faces possible criminal prosecution and financial ruin. Government agencies fear political fallout should the cleanup be unsuccessful. Contractors view the spill as a boon to their business and a way to recover capital invested in spill response equipment and readiness. Tourism officials worry about the effect on the island's income. Conservation groups demand protection of sensitive environments, while health officials fear harm to humans' health. Because of these competing interests and different viewpoints, an underlying lack of trust exists between each of these parties at a time when they most need to cooperate and work together.

The only apparent funding source for the *Berman* RP came from insurance provided by Water Quality Insurance Syndicate, the RP's insurer, with a spending cap of $10 million. The transition between RP and USCG control occurred a week after the spill at 6 a.m. January 14, when the USCG assumed full control and funding became a federal responsibility. From the outset the FOSC had access to and made use of funds from the Oil Spill Liability Trust Fund (OSLTF) for the government's response to the incident. Within days it became clear that the cleanup would cost much more than the RP's $10 million (ultimately almost nine times that amount by completion) and that the U.S. government would take over. MBI, the insurer's management contractor, was retained. A blanket contract was put into place with National Response Corp., the primary OSRO. Other contractors were kept on the job. The shift between RP and federally funded cleanup occurred in a supposedly "seamless transition."[8]

Still, this turnover was not without problems and elicited numerous comments such as, "Oh, no, not another Day One." The unique events of the *Berman* spill caused the confusion and chaos of "Day One" to recur with unwelcome frequency. For example, Day Two of the spill was "Day One" when a whole new cast of characters arrived on the scene. After the barge was sunk and another major oil release occurred, it was another "Day One." Each time USCG personnel were rotated out of key positions and replaced by "rookies," that was a "Day One."

Those who caused the accident did not slip away. The captain of the tugboat *Emily S* and his mate, the companies who owned the barge and tug, and the agent for the RP all came under criminal indictment by a Puerto Rico grand jury and were sued for civil damages by the commonwealth. The first mate surrendered his USCG license. Eventually, Bunker Group Puerto Rico, two other corporations and general manager Pedro Rivera were convicted on criminal charges by a U.S. District Court in San Juan, Puerto Rico, and face heavy fines and/or imprisonment. Rivera could face up to five years' imprisonment. Roy McMichael, Jr., the tug captain and his mate, Victor Martinez, pled guilty to criminal negligence and face up to one year imprisonment and heavy fines. The civil suit continues to the present.[9]

BRIDGING THE LANGUAGE BARRIER

Imagine the scene of the *Berman* spill with thousands of workers, representatives of 15 government and commonwealth agencies, and 50 contractors all speaking two very distinct languages, English and Spanish.

Felix Lopez, USFWS, regarded his major contribution as translator. In Puerto Rico this equates to more than simply explaining what the words of one language mean in another. It also means finding your way around an island where specific locations are not clearly marked, and directions to a spill site might involve following unmarked roads to turnoffs after "the last palm tree on the right."

When the spill spread to the northwest coast of Puerto Rico, oil threatened some very sensitive ecological sites in an area known as Shaks Beach near the small town of Aguadilla. Shaks Beach is hard to find. There are no sign posts. Crews had to act quickly to save sensitive turtle-nesting areas and manatee holes. Lopez gave directions typical of local lore.

> *"Follow the dirt road past the town. Look to the ocean and you will see all the surfers. Don't head to that spot. Instead, go up the hill a ways. When the road turns to the right, take the next dirt road you see to the left. Proceed until you run into the sea and a protected cove."*

No spill response coordinator from Savannah, GA could provide those directions. These kinds of directions were repeated hundreds of times daily and required many skilled local "translators."[10]

BUSINESS AS USUAL

USCG staff shortages became a problem. In addition to handling the *Berman* response and cleanup, the FOSC's staff were also responsible for handling normal USCG duties. At the time, these crises included responding to *mystery crude spills* off the east coast of Puerto Rico which affected six other U.S. Caribbean islands, a cruise ship grounding, a barge sinking at a dock, two chemical spills, and an out-of-service ferry stuck in San Juan Harbor.

The FOSC is both the Captain of the Port and the Officer in Charge of Marine Inspection. This means that normal duties of his office, such as prearranged inspections for the numerous ships calling at the port, must be handled on a routine and efficient basis.

Throughout the *Berman* crisis, the FOSC never had a stable staff of support personnel, due to the Coast Guard's temporary additional duty (TAD) personnel process. Individuals from other USCG units were assigned for temporary duty, and rotated in and out of the unit, requiring the FOSC to retrain key staff. The higher command seemed to treat *Berman* as a training opportunity rather than the crisis it really was. FOSC Ross eventually divided his permanent staff in half, assigning junior officers to routine tasks and keeping more senior officers on his spill management team.[11]

IMPENDING DISASTER

Berman was more than just another oil spill. The site of the grounding, the condition of the barge, and the season all threatened to turn a bad situation into a major economic and environmental disaster that might make the $87 million spent in cleanup look like a down payment.

January 7 was the height of the tourist season. The grounding took place close to San Antonio Channel, the "back door" of San Juan Harbor.

This vital harbor, designated a National Estuary, is home to many endangered species, such as the brown pelican and manatee.

Four to seven cruise ships entered San Juan Harbor daily. This meant that at any given time there might be 4,000 to 8,000 passengers and 12,000 to 24,000 pieces of luggage docked and discharged at the piers in San Juan Harbor. Hundreds of tourist shops eagerly awaited the dollars of happy visitors disgorged into the streets of Old San Juan. The FOSC could not shift the cruise ships to another Puerto Rican harbor. There are simply no other ports available which are large enough or deep enough to accommodate these ships.

Environmentally sensitive areas lay to the west of the barge, where manatees and pelicans make their homes and turtles return to lay their eggs in February and throughout the spring. The Mona Passage west of Puerto Rico is a passage area for thousands of humpback whales, who were due to arrive shortly. East and west of the site were sensitive mangrove and bird-nesting areas.

After the first inspection of the *Berman*, it was evident that the barge was damaged to the extent that it was a question of when, not if, she would break up. Until a sufficient quantity of the remaining oil on board could be pumped off and lightered to another craft, there was no hope of simply filling the barge with air and floating her away.

The FOSC faced a possible disaster of unprecedented proportion: spilled oil entering San Antonio Channel and closing the port in San Juan Harbor, trapping cruise ships and thousands of tourists with no alternative cruise ship facilities for those beginning their trips, and impacting the island's economy at the height of the season. There was also the terrible prospect of the barge sinking and becoming a permanent fixture at Punta Escambrón, with oil continuing to leak for years onto the shoreline, historic forts and ruins, and (no longer popular) tourist beaches.[12]

THE 36-HOUR FORTUNE 500 COMPANY

Capt. Ross described how he and his crew responded to the initial stages of the spill:

> *"If you take the number of people involved, the rate we spent money, and annualized that over the time of the cleanup, I'd have to say that in a short 36 hours we built a Fortune 500 corporation from scratch. We took 1,000 to 2,000 workers untrained in oil spill response, none of whom ever worked together, most of whom spoke two different languages, with spill fighting equipment coming onto the island from more than 1,500 miles away, and within 36 hours we were up and running at full force. Within two hours of the initial contact, we had a boom placed at the most critical spot, in front of the*

Dos Hermanos Bridge, to protect against a major disaster, oil flow-
ing into the San Antonio Channel and then into the San Juan
Harbor at the height of the tourist season. We beat the leading edge
of oil headed that way by about ten feet. "[13]

Capt. Ross accomplished this enormous task "one bite at a time" with
exceptional competence, and applied a true military Incident Command
System. He evaluated information received by the underwater dive team
and those working on board the barge. Based on the deteriorated state of
the barge and its position off Punta Escambrón, he considered the branches or possibilities of what could likely occur. He considered the most and
least likely branches, and the adverse consequences of each scenario. He
then examined the best preventive measures to follow in order to avoid
the worst-case scenarios.

Theory became reality through action. Capt. Ross and his staff identified equipment and manpower requirements and determined those that
could be filled immediately by local spill subcontractors. Available
resources were put to work as fast as possible. Calls were made to obtain
the people and equipment needed as quickly as distance allowed.
Decontamination sites were organized for cleaning equipment. Berth and
mess hall facilities were arranged for arriving crews.

Command headquarters was established at the USCG base. Base telephone technicians tapped into existing USCG telephone trunk lines to
establish necessary communication desks in the command center for each
agency. The Puerto Rico telephone company later wired in another 20 lines.

According to Capt. Ross, one of the most important aspects of dealing
with a spill of this magnitude is to focus on the logistics. "Dilettantes talk
only tactics and strategies. Military professionals talk logistics, logistics
and logistics," he said.

Confidence building became a major task for the FOSC and his staff.
Capt. Ross found himself in the unenviable position of assuring contrac-
tors, including NRC, that they would be paid from federal funds once the
RP's initial $10 million in insurance ran out.

Payment was not the only area where confidence building came into
play. The FOSC continually encouraged a high level of trust and communication between the various people charged with responding to the spill.
Ross recalled:

"We needed a coherent approach toward managing events. We all
had to be singing the same sheet of music. While making sure that
the team was pulling in the same direction, I couldn't run roughshod
over anyone. I consulted with the people in charge, got their exper-
tise and input, and made the final decisions."

Ross' skill and capability resulted in a large group of bureaucrats and business leaders who had never worked together before coming to a consensus of opinion and dealing with a major crisis. Rules and methods for handling first-time problems were developed along the way, and the FOSC kept a tight rein over this "organized chaos."[14]

The FOSC factored into his overall plan of attack for the spill certain unwritten elements. According to Ross, the *Berman* response strategy in its broadest outline consisted of the following elements:

Capt. Robert G. Ross, USCG (center). (Reprinted by permission of U.S. Coast Guard MSO, San Juan.)

1. Blocking San Antonio Channel.
2. Implementing other protection of resource strategies.
3. Securing the source through lightering and salvage.
4. Removal of gross contamination, floating and beached (later expanded to include sunken oil).
5. Final shoreline cleanup after steps 1 through 4 were accomplished.[15]

The *Berman* cleanup response was a success by any standard. Consider the spill itself, the potential for far greater damage, the absence of the RP, and the understaffed and ill-prepared responders. A brief analysis of the qualities that made the difference between failure and success in *Berman* also provides guidance for future spill response:

- Plan before a spill. Introduce all the players. Identify critical areas. Prioritize response. Divide up responsibility. Establish a management model. Train each participant in their role.
- Conduct lightening-fast initial response to the spill to minimize further impact and damage.
- Direct clear, competent evaluation of the circumstances, threats

and hazards for containment to prevent further damage. Spend minimal time in "storming" (consensus building) and focus on "performing" (doing the job at hand).

- Organize the troops.
- Select optimum, competent leadership.
- Form a competent command structure and plan to deal with the incident.
- Adapt the contingency plan model to the actual situation.
- Be flexible.
- Establish strong teamwork and communications. The working relationship between the FOSC and the RP should operate smoothly, with each party knowing its role and how to fit into the overall picture.
- Delegate duties and allow for the various viewpoints of those involved.
- Save and reduce to written manuals the lessons learned.
- Examine the spill causes to strengthen and improve prevention measures so a similar spill is far less likely in the future.

And, as Capt. Ross and others involved in the *Berman* response would no doubt add, having a little luck on your side helps, too.

SECTION II

CORAL: ROADKILL ON THE PETROHIGHWAY

THE CORAL REEF QUESTION

NO IMMUNITY FROM TIME – OR MAN

Nothing on our earth is immune from the passage of time, including the world's coral reefs. These reefs that ring the globe, teeming with abundant marine life, have become yet another victim of man's need for space, food, shelter, recreation and waste sites. For hundreds of millions of years, nature has bombarded reefs with storms, hurricanes and climate changes. Scientific study has determined that coral reefs have the capability to survive most natural disasters in some form or another and return to a healthy condition over time. The delicate reef systems and their inhabitants have not demonstrated the same tolerance to severe and repeated degradations from man. Man's entry into the world scene has tipped the balance – much more rapidly in recent decades – against the continued existence of reefs. Among man's more significant actions that harm and endanger coral reefs are:

- Development of coastal shorelines with resulting land degradation, deforestation and severe erosion.
- Clearing timber and mangrove forests with resulting sedimentation and loss of buffer zones between land and sea.
- Dredging and port development, with ports becoming sinks of fine sediments and pollutants.
- Over-fishing to feed increasing numbers of people.
- Discharge into the ocean of sewage, agricultural byproducts and industrial waste.
- Dumping directly or indirectly of land-based toxins into the water.
- Burning fossil fuels and releasing chemicals into the earth's atmosphere that many reef scientists contend damage the protective ozone layer, accelerate global warming and effect drastic changes on fragile reef ecosystems.

Even when people engage in supposedly non-harmful activities, they leave behind a wake of destruction. Man's quest for recreation, (diving, snorkeling, fishing and boating) all stress and damage the undersea world.

OUR OIL CONSUMPTIVE SOCIETY

While all threats that have a harmful impact on the world's coral reefs are worthy of our attention, this text will focus primarily on a most alarming issue: Oil Pollution. One oil spill can cause terrible harm and long-term devastation to a coral reef in a single event.

Modern man needs fossil fuels, particularly oil. Processed crude oil powers our cars and other vehicles, heats our homes, runs our factories, paves our roads, and provides the raw material for more than 3,000 everyday products.

Even those of us who make a conscientious effort to reduce dependency on our cars by car pooling or taking public transportation are still part of the problem. Daily we consume petroleum byproducts. Take a quick look around your apartment or house and count the number of synthetic items surrounding you. Start with the clothes on your back, the roof over your head, the faucet on your sink, maybe even the sink itself, the paint on your walls, the fiber in your carpet, even that watchband on your wrist. These are all petroleum-based products, everyday household items created from petrochemicals. There is no escaping the obvious fact: we are a society of oil consumers, and an enormously consumptive one at that. According to one source, residents of California, considered the most environmentally conscious state in the U.S., "… consume more petrol than any single nation in the world (except the USA)."[1]

Crude oil is a mixture of naturally occurring substances which can have measurable toxicity to living systems under most circumstances. Although there is some dispute regarding the level of threat oil presents to coral reefs, the toxicity of oil cannot be denied. We, the ultimate consumers, are responsible for the risks posed to these marine ecosystems from oil spills.

As long as we demand oil for our daily existence, economic reality dictates that someone will supply it. Petroleum and petroleum products are shipped to us from the discrete locations around the globe where this substance can be found or refined. The primary means of transporting oil is by large, ocean-going tankers and tank barges. Even with the best safeguards in place, millions of dollars spent in prevention, national and international programs, strict laws, regulations and criminal punishment, it is inevitable that vessels navigating the seas will have accidents. When accidents occur, typically thousands of gallons of oil flow from the breached holds of these huge ships directly into ocean waters.

SO MUCH ABOUT SO LITTLE

The current opinion is that oil spills at sea make up only five to ten percent of all oil-based pollution. The other 90 to 95 percent of oil-based pollution enters the water from land sources.[2] Surprisingly, one of the worst offending sources is the oil that drips innocently from your car onto the street, trickles into sewage drains which empty into pipes that eventually find their way to the sea.

If oil spills at sea constitute such a small fraction of overall oil pollution, why the big concern? Because oil spills that affect our oceans are not rare occurrences at all, and research shows that the effects of oil spills are particularly devastating to coral reef ecosystems. Small spills happen daily. The continued threat and occurrence of these spills, from the *Amoco Cadiz* (1978, Brittany Coast, France), *Exxon Valdez* (1989, Prince William Sound, Alaska) and *Braer* (1993, Shetland Islands) to the smaller daily events, create a crisis for our oceans. We can prevent these spills from happening in most cases, ending one source of human pollution of our seas.

Recreational divers, who make up one of the fastest-growing recreational sports in the U.S., together with snorkelers, have long enjoyed the beauty and exuberant marine life associated with coral reefs around the world. But suppose you have never seen an actual living coral reef. Why should you care what happens to them? What difference does it make to you whether coral reefs live or die?

In order to better appreciate the importance of coral reefs and the necessity for their survival, it is helpful to first learn answers to these questions:

- What is a coral reef?
- Where are these unique living structures found?
- What creatures make up the delicate balance of beings living in, upon, around and forming part of the reef?
- What natural and man-made conditions threaten these unique beings?
- How do coral reef systems benefit our world?

Armed with this vital information, we can more fully understand this spectacular element of the undersea world and grasp the serious threats posed to reef systems from oil spills. You do not have to be a marine biologist, a scientist or an oceanographer to understand coral reefs. All that is required is a love and appreciation for the natural world, which includes the underwater realm that makes up three-quarters of our planet, and about which we still know so little – and a healthy curiosity.

At the end of this voyage of discovery through the fascinating world of coral reefs, you may well experience a shift in attitude from "Who cares?" to "I care about coral reefs – what can I do to help protect them?"

THE CORAL REEF
AND ITS INHABITANTS

AN OASIS IN THE OCEAN

If you have never seen and experienced a coral reef firsthand as a diver or snorkeler, the vibrant reef world may seem as remote and unimaginable to you as a tropical rainforest. In fact, it is both as diverse and magical as a rainforest, and as exotic and foreign as another world. Perhaps you have glimpsed the wonders of coral reefs from underwater nature programs on television, or from magazine articles featuring magnificent underwater photography. It is likely that any exposure to the reef world left you hungering for more – maybe a vacation visit in person.

Whether you live on the plains of Kansas, in the mountains of Colorado, among the fertile fields of Iowa, or any number of places where a tropical beach and clear, blue ocean waters are far removed from your everyday experience, you are in the minority. You are among the billions of people who do not live near the water, but of the total world population of some 5.6 billion people, over three billion live near the world's oceans. This means that almost 60 percent of the human race lives within 60 kilometers of the *Coastal Zone*, that area of land bordering the ocean, the nearby lands directly beyond beachfront, and the river catchments where waters flow to the sea. You should not be surprised at these numbers. After all, water covers 70 percent of the earth's total surface.[1]

Coastal dwellers are on the increase. According to projections, by the year 2025 almost 8.5 billion people will reside near coastlines, with many of those residents living in developing countries in the tropics.[2]

Even if you live near the water, you might not be near a coral reef. The tropical ocean has been described as a vast "biological desert."[3] Coral reefs are the *oases* in this watery desert. Much of the earth's tropical seas are extremely poor in nutrients and have low productivity, much like a clear soup. The reefs have developed mechanisms to trap and cycle nutrients efficiently. They provide the chunks of meat and vegetables that flavor the broth.

Coral reefs are scattered throughout the ocean in specific places in the world. These ecosystems lie in latitudes that fall between the southern tip of Florida and mid-Australia in a band that runs 30 degrees north and south of the equator in warm subtropical or tropical oceans.

The stony corals that compose the reef prefer temperatures ranging generally between 25 to 29 degrees Celsius (75-85 degrees Fahrenheit) but can survive in temperatures between a low of 18 to 20 degrees Celsius and a high of 33 degrees Celsius.[4] The majority live only in shallow waters, at depths of no more than 30 meters (about 100 feet), although some hard corals can survive at greater depths. Over 100 countries share the distinction and privilege of being neighbors to coral reefs.[5]

CORAL CONDOS

The basic unit of a reef is the coral polyp, a tiny animal just a few millimeters across. Polyps grow in large aggregations, connected by common tissue, to form a colony. A colony may be up to a few meters (1 meter is about 40 inches) in size. Colonies in turn may group together to form reefs which can extend for tens or hundreds of kilometers (1 km is about .62 miles).

There are many types of polyps. The principal "mason" or builder of the reef community is an animal called the *stony coral*. Stony corals are less complicated to understand than your household dog or cat. The scientific name is *Scleractinia*. Corals derive their name from Greek, meaning "hollow gut," referring to the simple structure of the coral animal, the *polyp*. The polyp is composed of a central hollow tube or cylinder shape through which the coral feeds, breathes and removes its own waste. This tube is made up of a jelly-like tissue. The mouth of the tube is surrounded by tentacles which capture and bring food into the digestive cavity in the center of the animal. *Nematocysts* are stinging threads in specialized cells located along the tentacles. Stony corals capture their prey in these tentacles and then inject the victim with a poison-like substance.

The juvenile polyp, the planula larva, settles on a clean substrate and produces a rocky material composed principally of the chemical compound *calcium carbonate*. The stony coral deposits this hard waste sub-

CORAL POLYP

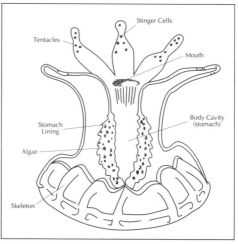

(Source: *Teacher's Guide to Coral Reef Teaching Kit*, produced by World Wildlife Fund and RARE, Inc. as part of the Caribbean Environmental Education Program. Reprinted with permission from Katherine Orr 1985, courtesy of World Wildlife Fund.)

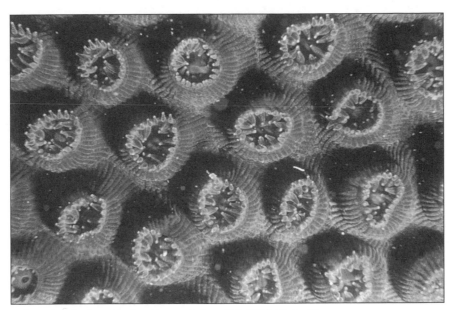

Colony of coral polyps. (Reprinted by permission of World Wildlife Fund.)

stance around the lower part of its body, creating a *corallite* cavity for itself. The coral polyp attaches to this cup with its muscular tissue, enabling its tentacles to reach out and grab prey drifting by. Most stony corals pull their bodies into this cup during the day for protection from coral-eating fish and other reef creatures that consume coral. At night the tentacles and hollow tube emerge to feed upon shrimp, larvae and plankton floating by in the sea.

Stony corals are tiny animals, ranging in size between 1/16th of an inch to 1.0 inches. Coral polyps are connected together by their living tissue like a limestone condominium. The result is mountains of coral produced by minuscule animals. The world's coral reefs cover approximately 600,000 square kilometers of the ocean's near surface. The largest and best known reef, the Great Barrier Reef in Australia, is a massive structure 2,000 kilometers (1,200 miles) long.

Stony corals are the primary reef builders. Unlike many other coral polyps, stony corals host an interior resident, a single-cell microscopic alga known as *Zooxanthella*. *Zooxanthellae* look nothing like your typical house plant, lacking roots, leaves and stems. Coral tissues host between one and two million algae cells per square centimeter.[6]

The polyp and algae team up to build reefs in this process. The coral animal secretes waste materials, such as carbon dioxide, nitrogen, ammonia and phosphorous from its breathing and digestive processes. The alga takes these waste materials and, using energy from sunlight, converts them into sugars, oxygen and carbon compounds in a process known as *photosynthesis*. (Common household plants use sunlight and the nutrients in their soil to accomplish much the same task.) The stony coral converts the raw materials produced by the alga, particularly oxygen, into nutrients for itself. Photosynthesis stimulates production of calcium carbonate by the polyp, which forms its skeletal cup. Exactly how photosynthesis enhances formation of reef building material is unknown, even after considerable study. What is clear is that a typical stony coral such as brain coral kept in the dark for long periods of time will stop depositing calcium carbonate. Sunlight generates the energy that makes the process work.[7]

The coral and the algae both benefit, coexisting in a *symbiotic relationship*. The coral receives food, oxygen and material needed to construct its reef home. The algae obtains a residence, with carbon dioxide and nourishment in the form of coral waste products such as nitrogen and phosphorous.

Coral reefs grow only in clear, shallow waters because the algae inside the coral needs sunlight for photosynthesis to occur, and the sun's rays cannot penetrate deep, murky waters to reach the algae deep inside the corals. The optimal condition for this successful symbiosis is clear water of limited depths in warm tropical or subtropical oceans. Only in the narrow band encircling the equator does this unique combination of depth, warmth and clarity exist.

There is more to the process of forming a complex coral reef beyond the relationship between stony coral and algae. Other partnerships are also at work. A coral reef can perhaps best be defined as " ... a submarine structure formed through the active growth of organisms and the cementation of carbonate skeletons by other organisms."[8] Essentially the reef is composed of layers. When a stony coral or another reef creature dies, its soft body disintegrates. The part that is not consumed by reef organisms, the hard skeleton, remains to form the foundation of the reef. Much like a house composed of bricks, the skeletons of dead coral pile up on top of each other over many years, with the live coral always forming the uppermost layer of this ancestral heap.

Another process, sedimentation, adds to the active coral skeleton deposition of the stony corals. The *sponge* contributes to the reef building process. Sponges, which may resemble a kitchen sponge or other more exotic forms, along with other reef creatures become part of the reef structure when they die, their own bodies creating yet more layers.

REEF CEMENT

As cement or mortar bind the bricks of a house, coral reefs need a binding element. Yet another partner, the *coralline alga*, acts as material to intertwine reef layers. These multicellular plants extract calcium carbonate from the sea and form calcite deposits throughout their cell walls.

They grow on top of the reef surface and cement over the top of coral and other substances. Sometimes this alga forms a pinkish-colored encrustation on the reef structure.

Another contributor to the reef sedimentation process is a bright-green alga called *halimeda*. These algae consist of calcified segments of flattened disks joined together by compartments in which calcium carbonate is deposited. When these leaflike forms break down, they form sediment and sand that fill in gaps in the skeletal structure. The powdery Caribbean beach sands owe their whiteness to, among other creatures, these tiny flakes of halimeda.

These two types of alga, red and green alga, the dead animal bodies of hard or stony corals, and some other reef creatures such as gorgonians and mollusks, join together to form one massive reef structure.[9]

While we consider ourselves participants in the age of "recycling," people are not nearly as accomplished at the task as this ancient underwater world. The reef superstructure is a true example of a self-sustaining system which uses every part of itself to produce life. The result is one immense formation of organisms joined in the process of dying while others spring to life. Corals, octocorals, hydrocorals, anemones, sponges, algae and other living organisms grow and colonize the outer layers of the reef. Underneath this mass of life are skeletons of dead coral, filled in by coralline algae, calcareous algae, sponge, mollusk and other reef sediment consisting of creature corpses. The outer layer of the reef grows upward toward the surface and then expands outward laterally in a slow and orderly pace.[10]

Coral garden. (Reprinted by permission of World Wildlife Fund.)

One lover of reefs has described the life and death process of the coral system in human terms: "In the sea there always seems to be this relationship of life forms. One form dies a little and another is born, like changing expressions on a human face."[11]

THE HITCHHIKER

Where does the coral come from to form a new reef? Coral reefs are produced through a process called *recruitment*. New members move into a coral community the same way new neighbors move into a neighborhood. Resident coral can also reproduce and create new organisms. Corals possess the remarkable ability to reproduce both sexually and asexually.

The beginning of this process is the forming of a new coral animal. *Brooding* and other corals produce eggs and sperm inside the polyp's soft body. The polyp releases the sperm, which is carried on the water currents until it is drawn into the mouth of another polyp and fertilizes the egg inside. When ready, the polyp releases the maturing *planula larva* into the sea.[12]

Yet other corals reproduce by *spawning* or *broadcasting*. This involves releasing massive clouds of sperm and eggs into the ocean where the sperm finds the egg and fertilizes it to produce the planula larva. By some wonderful mechanism the coral colonies of a given species release their eggs and sperm into the water at the same time. Observers of the Great Barrier Reef have captured incredible footage of the annual release of thousands of sex cells into the ocean simultaneously.

Caught in ocean currents, the planula may drift for days or weeks. For some types of coral larvae, the maximum life span is between 17 to 23 days.[13] Others live as long as 35 days.[14] Some voyagers attach themselves to floating objects such as coconuts, driftwood or volcanic pumice. These hitchhikers or free floaters may "raft" the ocean currents great distances. Scientists have found coral larvae growing on pumice in a coral reef in Hawaii that came from the east coast of Mexico.[15] Larvae from the Little Bahamas Bank are believed to travel about 15 days before reaching Bermuda some 1,100 miles away.[16]

In order to create new reef, the survivors eventually leave the ocean surface and drift down toward the bottom habitat in search of their new home. The conditions for that home must be perfect: a hard surface relatively clean of other organisms who might compete with the larva for space, and free of silt so the new polyp may establish firm footing.

HOW SLOWLY THE GARDEN GROWS

Once the planula larva finds its unique place, a phenomenal shift takes place. The planula begins to grow by a method called *budding*. The creature settles its soft body onto the structure and begins to form its distinctive corallite, protective cap. This polyp divides into an exact replica of itself. New polyps form inside or outside of the original coral animal. The daughter polyp produced outside the coral attaches to the parent by its living tissue. Or, in other corals, the parent cup enlarges to make room for the new polyp growing inside it. In this manner huge colonies of coral polyps form a reef.[17]

Another way in which reefs are created is by asexual reproduction through *fragmentation*. Hurricanes, heavy surf and reef boring organisms, which attach to the reef and consume coral, break corals apart or detach them from their bases. The survivors find new substrate, start the process of budding, and grow into a colony.

Assuming that all the right conditions exist, different types of coral colonies grow at different rates. Eugene Shinn, a geologist, diver and long-time marine scientist, explained how rapidly a staghorn coral colony can grow. Aptly named for its resemblance to the antler rack on a stag, this coral can gain four to five inches a year, and each stalk may branch into three others. One clump of staghorn coral made up of ten branches could produce 35 *miles* of branching coral within ten years.[18] Even under optimum conditions such rapid increase does not occur due to natural pruning. Storms, waves, the depth of the ocean, and reef creatures limit the increase in size of staghorn coral. In particular biologists have found a worm (the fireworm) which feeds on this coral and literally digests whole stalks.

Corals other than branching types grow at a much slower pace. Where staghorn coral might produce ten to 20 centimeters (about 8 inches) of skeletal material in one year, boulder coral (another basic reef builder) needs 20 years to form the same amount of reef colony.[19] Common star coral grows at less than one centimeter a year. Other coral animals grow at an even slower rate, only five to 25 millimeters (.2 to 1 inch) per year. Slow growers require hundreds of years to reach a significant size.

Shinn's growth rate figures apply to coral colony growth. Upward growth of the reef, through the growth of these colonies and accumulation of debris in the crevices, proceeds at a much slower rate. In fact, colony growth of individual species hardly translates into growth of the whole reef system. One scientist estimates that upward growth of the actual reef structure occurs at the phenomenally slow rate of .2 to .8 millimeters a year (roughly 1/100th to 3/100th of an inch.)[20]

These rates assume that only natural impacts affect the polyps, such as fish and other reef predators, storms and rising sea levels. Suffice it to say, hundreds and even thousands of years must pass before a reef realizes significant growth. On the Great Barrier Reef marine biologists have identified corals between 800 and 1000 years of age.[21] When man's often harmful activities are added to the equation, the time span for reef growth takes even longer – if it occurs at all.

MODERN CORAL

Coral reefs have been in existence in various forms for hundreds of millions of years. Although estimates differ, Caribbean reefs are relatively young, less than 10,000 years old.[22]

To put the history of reefs into proper perspective, consider this: the earth is 5000 million years old. Over 4000 million years ago the simplest form of life, one-celled animals, appeared. 2000 million years ago the first reef-like beings arrived. These creatures were blue-green algae, able to trap calcium carbonate and build limestone formations in mounds known as *stromalites*. The higher order of plants and animals did not come into existence until about 1000 million years ago. About 600 to 500 million years ago, during the Cambrian period, the first true reef building animals appeared. These early beings were soft-bodied, sponge-like creatures growing amid algae. Stony corals first appeared 400 million years ago. Finally some 60 to 65 million years ago the modern reef as we know it formed.[23]

Reefs are naturally hearty and resilient. Throughout millions of years, these complex beings have withstood changes in temperature and sea level, only to return in new forms, shapes and places on earth. Reefs have changed continuously since their beginning. Great natural upheavals during hundreds of millions of years have changed their make up and distribution. The last such cataclysmic event took place during the Ice Age 1.8 million years ago when the dinosaurs died out, as did two-thirds of the world's underwater structures. The seas fell, leaving most reefs high and dry.[24] In spite of all that nature has thrown at them, from shifting land masses over the earth's surface to tremendous climate changes, coral reefs have endured to face their greatest challenge – man.

THE WEB

Coral reefs are a living example of one of the most refined systems of checks and balances found in any natural system. The peaceful coexistence of the alga within the coral polyp is only one form of mutual sharing that takes place in this complex world. Almost every creature in this underwater kingdom depends on its neighbors for some form of service, from the lowliest microscopic plant or animal to man, the *ultimate consumer.*

The reef demonstrates in microcosm the greater *web* in which we all live. Each animal and plant provides a food source for increasingly larger predators all the way up the chain to humans. At the bottom of the food chain are *plankton*, meaning "drifters" or "wanderers." These essential organisms are both plants, *phytoplanktons*, and animals, *zooplankton*. They cannot control their direction by swimming, but rather are swept along by currents.

Zooplankton may be juvenile animals living in a temporary phase before they attach to the bottom and develop into adults, like the coral polyps, or they may be organisms that spend their whole lives drifting. Each of these tiny creatures is at the base of the larger food chain, consumed by the coral polyp, by other one-celled organisms, by sponges and by juvenile or small fish living on the reef. These smaller beings are then consumed by larger reef creatures. The jellyfish and manta ray feed upon zooplankton. The butterfly fish devours live coral tissue. The damselfish "farms" seaweed. Hawksbill turtles digest sponge. Parrotfish, sea urchins and other reef grazers mow down the algae, creating an even "turf" and leaving space to recruit new coral polyps. These sea creatures ultimately become prey for the larger reef predators: groupers, sharks and finally humans.

The coral reef is the ultimate "recycling center." Every plant and animal, regardless of its size, whether microscopic or easily visible to the eye, lives and dies on the reef, becoming nourishment for others. In this manner the limited nutrients of the reef waters are used over and over again in a system that works exceptionally well if left undisturbed. Reefs can adapt to various stressors because of the existence of redundant, or *backup* species, so that the reefs can cope, in some cases, with the loss of one species out of the entire reef pool. However, certain of the species making up the reef, *keystone* species, are extraordinarily important to the entire ecosystem. Stressors that affect these species have an impact on the whole reef ecosystem. One keystone link or organism removed from the chain affects the whole network.[25] Man is part of this system. We depend on the reef, just as our actions may affect coral's eventual viability. We are nourished from the reef's byproducts.

CHECKS AND BALANCES

This food chain is only one form of the interdependent relationships existing in the reef environment. More exotic living arrangements take place between reef creatures. The sea anemone and the clownfish enact one such colorful relationship. The sea anemone, a relative to coral, is a colorful animal with numerous tentacles surrounding its mouth. Usually an anemone lives in splendid isolation from other anemones, attaching itself to a coral or a rocky outcropping on the reef.

Most fish steer clear of the anemone because of the lethal stinging cells along its tentacles. In some manner over time, the small clownfish has developed resistance to the anemone's sting and can reside comfortably within its waving tentacles. The current theory holds that the clownfish coats its body with a mucus it secretes and thus protects itself against the anemone's sting. The clownfish finds shelter and safety within the anemone, avoiding the jaws of larger fish. The anemone receives a valuable benefit as well. The colorful, striped clownfish lures other fish into the anemone's deadly tentacles. The anemone also consumes waste products from the clownfish.

An even stranger mutually cooperative relationship exists between the smallest of creatures, the cleaner shrimp or tiny wrasse, and some of the largest reef fish. Over 50 species of cleaners survive in happy coexistence with their natural predators by providing a "valet service" for their larger neighbors. The big fish swim to specific cleaning station locations where they signal to the smaller cleaning fish or shrimp that they are ready for grooming. Fish may line up at these cleaning stations like autos at a car wash. Groupers, moray eels and other reef dwellers open their gaping jaws and allow the undaunted tiny fish and shrimp to crawl inside and perform their work. These cleaners feed on parasites that grow on the skin or in the mouth of the fish. So important is this relationship that biologists predict the decline of overall fish populations without these "fish washes."[26]

REEFS ARE NOT ALONE

Reefs are part of the coastal world, and while they have qualities of an independent, self-contained system, they rarely exist in isolation from the other important elements of the coastal ecosystem. Often when you swim from the beach to a reef, you start from a white sand beach, broken or occupied in part by *mangroves*. These woody plant communities fringe about 25 percent of the world's coastlines. More than 70 species of mangroves exist worldwide within the tropics. They are invaluable to the marine world's good health, sustenance and survival.

Mangroves provide critical breeding ground for juvenile fish and crustaceans of all sorts, oysters, mussels, snails and other sea creatures that attach to or shelter in their roots or limbs. They act as a sink by filtering fine sediments before they reach the ocean and coral reefs. These dense, strange-looking plant communities, whose root systems appear to stand awkwardly above the water, contribute organic matter (twigs, leaves, wood, bark and the bodies of associated animals) to the water body. Another important role of mangroves is that they provide a safe habitat for many animals. Also, the complex root system stabilizes the shoreline.[27]

Seagrass beds are flowering plants with long blade-like leaves that live underwater. Seagrass grows from an underground root called a rhizome, which anchors the grasses to the sea floor. There are approximately 45 species of marine seagrasses in the world. In the marine ecosystem these fields of plant material function as important nurseries for juvenile fish, conch, snails and other reef larvae. They act as a natural sieve, trapping and settling land sediment and keeping the waters clear. A seagrass bed may be alive with tiny fish gliding by, their silver-grey bodies no bigger

Coral Garden. (Reprinted by permission of World Wildlife Fund.)

than your pinky finger, usually in a frenzy of activity. These beds provide grazing areas for reef herbivores, turtles and other animals that depend on seagrasses as food source. They stabilize the sea floor, reducing beach erosion and stress to other marine habitats.[28]

The coral reef benefits its mangrove and seagrass neighbors. The reef superstructure shelters the two coastal systems from erosion and the impact of waves, contributes plant and animal material as food and nutrition, and builds up the sandy bottom habitats in which the mangroves and seagrass thrive. These small worlds are linked like a series of interconnected webs that together form a larger web system. The ill health, degradation or outright destruction of one element directly threatens the others. When scientists determine the health of the world's reefs, their assessment includes the condition of neighboring seagrass beds and mangroves.

AN ALIEN WORLD

Nowhere else on earth, except perhaps in rainforests, can one view such a vast array of exotic individuals as is found in coral reefs. In fact, reefs have often been referred to as the "rainforests of the ocean."[29] Some scientists believe them to be equally as rich in diversity as their better known counterpart.

The very names and appearance of reef inhabitants suggest the fanciful nature of this world that stirs the imagination with dream-like images of cartoon animation and science fiction. On the bottom of the sandy floor, you can find sea cucumbers, slow-moving dark-bodied slugs, tentacled sea anemones, spiny sea urchins, marine snails and reef mollusks. Near the shores of the beach you might spot many-armed sea stars, coral crabs, long-snouted gars, spiny lobsters and abalone

Corals sport delightfully descriptive names: brain, star, elkhorn, staghorn, flower, tube, rose, leaf, ribbon, saucer, fungus, pencil, cactus, pillar, finger and more. Soft corals bear equally visual names: sea whips, sea plumes, sea fans, sea rods, sea pen, snail fur. Other creatures abound, sea anemones, jellyfish, sealillies, feather stars, basket stars, sand dollars, sea squirts, lancelots, tube worms, light bulb tunicate, and the nudibranch. Reef fish include butterfly, clownfish, trumpetfish, gars, groupers, snappers, wrasse, gobies, surgeonfish, parrot fish, soldier fish, squirrel fish, and flashlight fish.

Consider our need to name creatures of the underwater reef world after familiar objects associated with our land-based existence. One might describe this tendency in terms of our anthropomorphic proclivities, or a host of other theories. But perhaps this naming simply reflects our fascination with this alluring alien world, and our need and desire to draw closer to it by creating an illusion of familiarity.

NATURAL THREATS TO CORAL REEFS

Coral reefs do not enjoy an easy life. They are at risk from natural and man-made causes. As one marine biologist explained, "Whatever happens on the land or air affects the water and reefs below."[1]

EYE OF THE STORM

Hurricanes and tropical storms wreak havoc on reefs. Gene Shinn, a renowned coral expert, believes hurricanes can cause more damage to reef structures than almost any other natural or man-made force. Hurricane Hugo, which struck the Caribbean in September 1989, generated winds speeds between 220 to 274 kilometers per hour, and produced waves exceeding 40 feet.

Such force can literally break the reef into fragments. Miraculously, these structures seem to recover from hurricane destruction. Shinn and other scientists have studied the effect of hurricanes on reefs. Reefs grow back as the surviving organisms heal or as new polyps bud from remaining corals. Storms can bring living coral into new areas. If the coral finds clean substrate free of algae, it can attach, form a limestone base, and a new reef begins.

Shinn estimates staghorn coral reefs damaged by hurricanes nearly recover within five years. According to Shinn, hurricanes have destroyed or seriously devastated the Florida Keys reefs once every six years for the past 10,000 years and still they survive.[2]

Caroline Rogers, a highly regarded marine biologist at Virgin Islands National Park, is more guarded in her opinion. Fast growth of even the branching corals assumes the existence of a healthy reef system. Coral polyps need unoccupied hard substrate, free of macroscopic algae, with a plentiful supply of herbivorous fish to graze algae and produce clean substrates for the stony corals, in order for reefs to regenerate. Not all corals are as quick to recover as the branching variety. Slower growing species such as head corals can take several decades to return to their pre-hurricane condition.[3]

Even if the reef recovers, scientists disagree on what the future reef may look like. Recovery may not necessarily mean that a reef returns to its former state. The cover, species number and diversity may differ before and after the hurricane. Some biologists believe there may be a loss in the types and numbers of coral creatures formerly inhabiting the reef, particularly where there are too many hurricanes, too much damage, and man-made impacts. The reefs near Jamaica are examples of what happens when all these forces coexist or occur sequentially.[4]

ALGAE EATERS - THE HUNGRY SEA URCHIN

Whenever the delicate check and balance system of the reef is upset, the superstructure suffers. For unknown reasons, certain primary algae grazers such as the sea urchin disappeared from Caribbean reefs in the '80s (although sea urchins are now making a comeback after the 1983 die-off in parts of the Caribbean.) One researcher studied what happened to Jamaican reefs when 99 percent of the population of *Diadema antillarum* (sea urchins) died in two years between 1982 and 1984. This die-off was preceded by two major hurricanes which damaged the balance of the reef. Without these sea urchin "bulldozers" to actively clean the coral and make way for new polyps, *fleshy macroalgae* took over. Unlike the coralline alga that cements the reef structure, this macroalga grows at a much faster rate than the coral polyps. Because of its rapid growth, fleshy alga soon takes over new coral. In Jamaica, people made a bad situation worse by over-fishing such herbivores as parrotfish, surgeonfish, and others, causing devastation to Jamaica's reefs. Formerly, these structures were home to 60 species of reef-building corals. Over a 15-year period, the balance shifted and coral cover declined from 52 percent to 3 percent, while algal growth increased from 4 percent to 92 percent.[5]

KILLER ALGA

Alga has staged an unprecedented attack on coral which scientists believe may transform once viable reefs into undersea deserts. Two diseases deadly to reefs are *Black Band* and *White Band* disease. Black band disease is believed to be caused by a blue green algae that produces a black mat which kills living coral tissue, leaving lifeless skeletal material, and has been known to deplete an entire colony of coral within one season. The cause of white band disease is unclear, but this disease kills coral tissue, leaving behind bare coral skeleton which rapidly becomes home to algae.

Black band disease. (Reprinted by permission of National Park Service.)

Biologists suspect man's actions may be a contributing factor. Alga performs an important role for the reef, acting as cement for the skeletal structure and a food source in the nutrient web. Reef grazers such as sea urchins and fish control the growth of this single-celled plant by cropping it. Man upsets this balance in two ways:

- Pouring tons of treated and untreated sewage, fertilizer, chemicals and other nutrients into the ocean converts nature's nutrient-poor environment into a harmful and chemically rich one. Fleshy algae thrive in an environment filled with phosphates, nitrates and other compounds; coral does not. While alga is good for the reef, too much fleshy alga becomes a problem.
- Over-fishing our reefs (which can occur easily without restrictions, due to the abundance of fish in shallow depths close to shore areas) removes the alga's natural predators.

Some scientists believe the end result of this set of conditions may be to cause these deadly diseases.[6]

CROWN OF THORNS

The presence, not absence, of another reef animal has caused consider-able damage to certain reefs in the Red Sea, Southeast Asia and even the Great Barrier Reef. The *crown of thorns* starfish consumes coral by pushing its stomach out through its mouth and attaching to the coral. The starfish secretes digestive juices that turn the live coral into an edible soupy sub-stance it can absorb. This consumptive process attracts other starfish, caus-ing a grotesque feeding frenzy much like sharks circling bleeding prey.

Again, humans may be partially responsible for this natural disaster. Our demand for housing and space has led to deforestation and destruc-tion of coastal mangroves. When heavy rains fall, sediments and rich nutrients from eroded soil flow into the ocean, magnifying the number of phytoplankton in the shallow waters near the reef. Starfish feed on these tiny organisms and may be attracted into nutrient-rich reefs in greater numbers. Coupled with over-fishing and collection of many fish and other creatures which normally eat this harmful starfish, the crown of thorns ascends in the hierarchy and soon rules the reef.[7]

CANARY IN THE COAL MINE

Variations in world temperature are part of the natural cycle. For mil-lions of years reefs have withstood global temperature change. Today earth's atmosphere appears to be heating up at an unprecedented rate, increasing faster than at any time during the past 6,000 years, with possibly disastrous consequences for undersea life. Since the mid-19th century, the earth's air layers have warmed up due to a phenomenon called the *Greenhouse Effect*, in which rays of sunlight bounce off the earth's surface in long-wave radiant heat. Greenhouse gases such as carbon dioxide, methane, nitrous oxide and chlorofluorocarbons (CFCs) in the global atmosphere trap this heat. Destruction of the stratospheric ozone layer by CFCs may create significant increases in the ultraviolet-B (UVB) radiation levels with increased penetration of harmful UVB into the water column.

Man contributes to the greenhouse effect by burning fossil fuels that run our cars, factories and homes. This process releases increasing levels of carbon dioxide. At the same time, we are destroying vast forests of trees and plants which would otherwise absorb much of this greenhouse gas from the atmosphere.

The greenhouse effect potentially threatens reefs by increasing ocean temperatures. The Intergovernmental Panel on Climate Change predicts a rise of 1 degree Celsius in global temperature by the year 2030, with carbon dioxide levels doubling by the middle of the next century (3 degrees C. by the year 2100.)[8]

This may spell disaster for coral reefs. They survive in a limited temperature range in which a difference of only 1 degree C. above average temperatures may have far-reaching consequences. One marine biologist describes this intolerance: "Tropical species like corals that are at or near physiological temperature limitation may be highly susceptible to even marginally developed temperatures."[9] A doubling of carbon dioxide in the atmosphere may lead to possible irreparable damage to the young corals in their planktonic larval stages, with a corresponding reduction in the ability of stony corals to produce calcium carbonate skeletons and build reefs.[10]

Coral bleaching may be an indication that coral reefs are not adapting well to accelerated climate change. Biologists detected coral bleaching and high mortality of coral reefs in the '80s and '90s in the Pacific, Costa Rica, the Galapagos, Florida, Puerto Rico, Jamaica and other parts of the Caribbean, and in Thailand and French Polynesia. Instead of finding healthy, vibrant reefs, divers encountered entire sections of bone-white coral.

Bleached coral. (Reprinted by permission of National Park Service.)

Biologists explain bleaching in two ways:

- Stony coral expels its friendly alga, the zooxanthella.
- The zooxanthella inside the coral polyp loses its pigmentation (coloring), leaving behind pale coral tissue.

If the bleaching event is brief, the coral may regain coloration (rebuild its zooxanthellae population,) as algae left behind in the coral divide and proliferate. The colonial animals can regain their color within a few months. Even so, rehabilitation is far from complete. Long-term recovery of normal growth and reproduction may take up to a year or more.

When bleaching is severe, most of the corals may die. Without its alga, the coral polyp cannot process sunlight through photosynthesis, the system by which the coral produces 60 percent of its building blocks – carbon compounds. Without this plant, stony corals lose their ability to form reef structure.

In bygone mining eras, coal miners carried a caged canary down into the shafts to warn them if poisonous gases were present. If the bird died, miners knew the shaft was unsafe and made their escape. Bleaching suggests that coral reefs may well be a "canary in the coal mine" for our age, serving as an initial indicator of destructive global warming trends that are more than reefs can endure.[11] According to several experts:

> "The condition of many of the world's coral reefs has reached a crisis point ... Global climate change may directly impose new stresses on reefs, or it may interact synergistically with other more direct human pressures to cause added and accelerated environmental damage. These (climate change) effects could accelerate the current rate of coral reef degradation in areas already stressed."[12]

MAN-MADE THREATS
TO CORAL REEFS

MAN, THE DESTROYER

The human race recreates everything in its own image, most especially things natural. If the environment does not suit, man changes it. Playing this maxim out to its fullest yields the *Four D's* of coral reefs:

- Damage.
- Degradation.
- Depletion.
- Destruction.[1]

In our efforts to *terra form* the earth, we dredge harbors and dig canals through coral reefs, cut down forests and burn away mangroves, clear off sand dunes and erode large areas of coastline. This technological onslaught causes nutrient-rich soil and sediment to flow into the sea. To this indirect contamination we add garbage to our oceans from direct sources such as oil tankers, raw sewage outlets and industrial plant waste pipes. Pollution from *nonpoint* sources such as chemicals, nutrients, soil, and whatever else might be lying on the ground in parking lots, lawns and city streets, flows through our forests, into our rivers, and eventually into our seas.

Marine *pollution* is "the introduction by man of substances into the marine environment, resulting in deleterious effects such as harm to living resources, hazards to human health, hindrance to marine activities including fishing, impairment of quality for use of seawater and related amenities."[2]

Man's actions combine with natural processes to damage reefs. Man-made stresses on and outright destruction of coral reefs can be grouped into five categories:

- *Mechanical and agricultural impacts* that release pesticides, promote soil erosion and result in the transport of nutrients and sediments into the sea.

- *Waste discharge* from industrial or sewage treatment facilities dumping chemicals and raw or treated sewage.

- *Over-fishing*, mining, and aquarium trade collection of creatures from the reefs, causing decline in important reef species.

- *Recreational use damage* from boat groundings, irresponsible snorkelers and divers, and anchoring on reefs. Gigantic cruise ships dispose of garbage within range of reefs, and repeatedly stir up clouds of sand that sift over reefs as they enter and depart from nearby docks.

- *Oil pollution* from catastrophic tanker spills and discharges, releasing toxic substances onto reefs and neighboring mangrove forests and seagrass beds.

SMOTHERED TO DEATH

The sea retaliates against man's destructive actions. In overdeveloped communities such as Miami, the Condado area in Puerto Rico, and most Caribbean island downtown seafronts, the beach erodes and sloughs off into the ocean. Ultimately the reef suffers. Upland silt and soil travel unimpeded directly into the ocean, carrying field pesticides and the flotsam and jetsam found along its way. Toxins include metals, phosphates, nitrogen, pesticides, chemicals, oil and salts, which poison coral.[3]

Sediment is harmful to coral polyps. Too much sediment literally smothers the coral polyps, impairing vital functions. Reduction in water transparency and light penetration by sediments impairs photosynthesis by the symbiotic alga within the tissues of the polyps.[4] According to World Resources Institute 1992 estimates, long-term deforestation occurs at the rate of 1 to 2 percent per year in many tropical countries with coral reefs. The result is a decrease in coral coverage from sediment flowing into offshore reefs located close to cleared land.[5]

CHRONIC EXPLOITATION

Coral reefs are a nursery for juvenile fish. Many commercial fisheries worldwide find their food source in reef fish. Modern man's increased skills are tipping the scale. Fish and other edible reef creatures such as octopus, squid, lobster and crab are caught in an infinite variety of ways, including nets, traps, explosives, spear guns and bleach. The larger members of species are killed in increasing quantities, leaving behind only smaller fish and other reef organisms. These, too, are hunted mercilessly. According to one study, over half the fish caught in Jamaica's reefs today are below minimum reproductive size.[6]

Experts from 100 nations gathered in Rome for a biennial meeting of the U.N. Food and Agriculture Organization concerning fish and fish stocks and concluded, "Virtually all species show signs of chronic over-fishing, from the large slow-growing fish to the small, fast-growing ones ... 70 percent of conventional species are fully exploited, over-exploited, depleted or in the process of rebuilding from over-exploitation."[7]

Food consumption is not the only problem. Vast quantities of exotic fish and reef animals are captured in the growing aquarium trade. While most states and countries ban taking coral from reefs, particularly through the Convention on International Trade in Endangered Species of Wild Flora and Fauna (CITES), removal of live coral and *live rock* is actively practiced by the ignorant and the greedy. By federal law and as a signatory to CITES, the U.S. prohibits the export of coral and coral products. Ironically, this law does not stop the import of coral and shells. According to one source, "In 1992, the United States imported about 85 percent of the raw hard corals and 98 percent of the live hard corals entering trade that year."[8]

This practice continues today.

BEATEN TO DEATH

In the chilly winter months, many North Americans escape to the clear blue waters and fantastic reefs of warm Caribbean islands:

- By 1995, the Caribbean region can expect to see more than 160 cruise liners operating in her waters.[9]
- In Virgin Islands National Park on St. John, tourist and resident use has tripled since 1976, drawing an estimated 900,000 to 1 million tourists per year.[10]
- Neighboring St. Thomas boasted a total of 1,745,000 tourists in 1993.[11]
- The Caribbean is considered the leading cruise ship destination in the world, according to Cruise Lines International Association.[12]

Tourism is big business for local economies, and *marine tourism* represents one of the fastest growing sources of foreign currency in certain parts of the world. This specialized market segment generates 60 percent of the annual gross national product for Turks and Caicos (for all tourism including construction), and $1 billion a year from tourism and commercial fishing in Australia.[13] Marine tourism produces $1.6 billion annually for Florida, and $286 million a year is generated from divers alone for the Caribbean and Hawaii.[14]

But tourists wreak havoc on reefs, too. Explained one scientist, "In reality, we are chronically beating our reefs to death in local, highly publicized places."[15] From cruise ships and passenger ferries to fishing boats, dive boats and sailboats, vessels run aground or carelessly drop anchor and chain on or near coral reefs. Steel and metal pulverize hard coral, tear sponges and soft corals to pieces, smash small patches of isolated reef, flip over coral heads, and can leave an entire reef system scarred beyond recognition. Recovery depends on recruiting new planula larvae or on asexual reproduction of remaining coral fragments. Such recovery, where the reef is not devastated beyond repair, may take upwards of 50 years or more.

Two examples illustrate the extent of boat anchor damage to reefs. In the supposedly protected waters of the U.S. Virgin Islands National Park, the 440-foot cruise ship *Wind Spirit* received permission to anchor in a bay off St. John in October 1988. The crew, through ignorance, negligence or stupidity, dropped anchor directly onto a reef. They attempted to remedy the mistake by hauling up the anchor and chain, but instead dragged them both along the reef, destroying almost 300 square yards of coral cover in the process. The U.S. Park Service brought a lawsuit against the ship's owners. At that time there were no federal regulations establishing a process to value the loss of such an ecosystem (unlike today's guidelines). Reconstructing coral reef loss proved difficult. Seven years after the anchor incident, the judge finally issued a ruling and the parties settled for $300,000.[16]

Another startling event took place in February 1985, when the 3,658 gross ton passenger ferry *A. Regina* ran aground on a reef near the Mona Island preserve off the coast of Puerto Rico. Gilberto Cintron-Molero, a longtime Puerto Rico resident and reef expert, studied this incident. According to Cintron, the ferry grounding provided a lesson in politics and futility. The insurance company and owners declared the ferry a total loss. At the time of the grounding, the ferry had about 210 tons of diesel fuel aboard, used to power the vessel. The USCG and others cleaned up the oil which spilled into the ocean and removed about 183 tons in total. Once the immediate emergency was addressed, no one seemed to know or care what to do with the ferry, which remained aground on the reef until late 1987, when the *A. Regina* was dismembered on site and the pieces were barged away to the Dominican Republic to be sold as scrap metal.

Leaving a vessel this size on the reef for so long caused massive reef destruction. Coral colonies were ground to rubble. The waters became filled with sediment from the constant grinding of the boat over the reef and stirring up the ocean bottom. The smashed reef structure no longer provided a suitable substrate for new coral to settle. Nearly three years after the incident, Cintron found no evidence of coral recovery, and no new coral colonies.[17]

These two incidents are not unique. Coral destruction from boat groundings and improper placement of chains and anchors are daily events. In 1984, the *M/V Wellwood* grounded on Molasses Reef, Florida, devastating 1,285 square meters of reef. The result was described as a "graded roadbed covered with a veneer of coralline debris."[18]

Snorkelers and divers can harm unsuspecting reefs simply by touch.[19] In their enthusiasm to view and photograph colorful reef sites, divers have been known to stand on coral, grab outcroppings, snap off fragments of these living creatures, or thrash about too close to fragile coral. Touching the coral surfaces can abrade, scrape and scour the fragile coral tissues, leaving the lesions on the coral susceptible to disease, particularly where the environment is marginal at best. Some solutions to diver destruction:

- Better diver education and awareness about marine and reef ecology during certification courses.
- Reef protection reminders by dive masters during dive briefings.
- Better enforcement against dive operators who chronically violate reef protection guidelines.
- Insist on diving without gloves to reduce the temptation to touch *Do not touch the living coral reef – doing so damages them!*
- Better buoyancy control to prevent clumsy, unintentional reef damage. *Don't be a Reef Wrecker!*
- Be a responsible diver, discuss reef preservation with your dive buddies, and support local reef preservation efforts where you dive.

UNPLANNED PLANETARY EXPERIMENT

Biologist Caroline Rogers of Virgin Islands National Park eloquently summed up the plight of reefs: "You can't see them, so it's hard for people to worry about what's happening to the reefs. They are not like the

Snorkelers exploring coral reef. (Reprinted by permission of World Wildlife Fund.)

redwood forests. If you bulldoze those down, everyone notices. With reefs, it's 'out of sight, out of mind.'"[20]

A sampling of observations and opinions from biologists, oceanographers, marine managers and scientists specializing in problems facing our oceans underscores the stressful effects of human and natural forces on coral reefs:

Aileen Velazco-Dominguez, marine scientist with Puerto Rico's Department of Natural and Environmental Resources: *"Contamination is never local. Whatever happens in one place in the water is imprinted and travels worldwide with the currents. Pollution at the base of the food chain moves upwards to reach man. For example, a spiny lobster spawns in Puerto Rico. Somewhere in its travels, the planula larva consumes polluted plankton. It spends the next six to 12 months adrift in the ocean's currents and ends up in Jamaica, thousands of miles away, as a full-grown lobster, caught by a local fisherman, and served up for dinner."*[21]

Evelyn Wilcox, former marine coordinator with World Wildlife Fund: *"We don't know how serious the problem of pollution and degradation of the marine environment is, but we do have the evidence that we are losing reefs annually due to climate change, marine pollution, and direct damage by recreational activities."*[22]

Terrence Hughes, marine biologist: *"It is highly probable that global reef growth is currently being out paced by reef degradation, with unknown consequences for the future."*[23] Man's interference is overtaking the adaptive ability of corals to negative conditions.

"In four of the well-studied sites, there are significant declines in coral cover and fish populations." These scientists predicted worsening conditions for the Caribbean because these reefs are "relatively small, and almost all are near islands with large populations and extensive tourism." They foresee only more of the same problems: hurricanes, loss of grazing sea urchins, over-fishing, increased macroalgae, sedimentation from local runoffs or dredging, and rising tourism.[24]

Carlos Goenega was a revered reef expert in Puerto Rico until his recent death. His comprehensive study, *The State of Puerto Rican Coral: An Aid to Managers* (1991), contains a complete evaluation of the coastal waters of Puerto Rico and its reefs. What he discovered was distressing: sparse coral coverage, thick sediment near the base of river outlets causing a decline of up to 80 percent of century-old corals, and poorly developed and stressed colonies of coral and other reef creatures elsewhere on all coasts. Man was responsible for most stress factors, including inland sedimentation, chemical pollutants, careless commercial and scientific collection of coral, and a high incidence of dead or dying reefs near overdeveloped coasts.

Goenega concluded sadly: "The available evidence suggests that coral communities may recover from natural disasters after several decades, but are likely to suffer irrevocable changes from man-made disturbances."[25]

Eugene Shinn predicts that if Florida's reefs continue to decline at present rates, many of these reefs will be "dead" in 20 years.[26]

The perilous state of Puerto Rican reefs is echoed throughout the Caribbean and Pacific, where limited species diversity and shallow average water depth render reefs at greater risk than their counterparts elsewhere in the world. But these reefs are not alone in their fragility. Two reef experts summarized a worldwide survey and announced disturbing news: "Reef damage from *anthropogenic* (man-made) environmental degradation is widespread, represents a much greater threat than climate change in the near future, and can reinforce negative effects of climate change." These scientists termed human interference an "Unplanned Planetary Experiment" and concluded that *"The human race appears to be in the process of conducting an unplanned planetary experiment in what might be loosely termed 'accelerated evolution.'"*[27]

GLOBAL CONCERNS

Global-level conclusions convey the experts' concerns:

Caroline Rodgers: *"There is no doubt that reefs in a number of parts of the world are facing serious degradation. But there are so many areas we don't know about, it's difficult to make global predictions. The resiliency of nature is amazing. However, resiliency works only up to a point. I am alarmed by what's happening. We know enough now to know that we need to protect reefs.* We just don't have the political will to do so (emphasis added)."[28]

Wilkinson and Buddemeir: *"Many of the world's coral reefs are severely degraded or destroyed by human damage to the environment, such as sewage and other pollution and sedimentation, and by direct over-exploitation and physical destruction; these impacts will increase with population growth and rapid economic development ... Reef communities are not well adapted to the combination of chronic and acute human (anthropogenic) stress and climate change, and their short-term survival is threatened by these stresses acting together (synergy), even though reef communities have persisted in the natural environment over evolutionary time scales."*[29]

Reef science statistics are alarming:

- Coral reefs have declined worldwide.
- Ten percent of reefs worldwide are considered beyond recovery. [30]
- Another 30 percent of the world's reefs face ecological collapse within the next 20 years.[31]
- More than two-thirds of the world's reefs will be considerably degraded within 20 to 40 years (60 percent over the next 40 years).[32]

- At greatest risk are reefs in South and Southeast Asia, East Africa and the Caribbean.[33]

There is cause for hope, and there are solutions to many of these problems. The efforts of skilled and devoted people the world over are creating positive change, especially in man's attitude toward this marvelous ecosystem.

Consider the status of oil pollution from marine casualties. The effects of these incidents on marine ecosystems, and coral reefs in particular, are serious and long-term. The U.S. and the world community are undertaking a series of strategies to prevent oil spills, and respond to them more quickly and effectively when they do occur. The same degree of intense scrutiny the shipping industry and its regulators apply to oil discharges must be initiated in dealing with other sources of marine pollution. The fate of our reefs is at stake.

OIL: THE UNWANTED SUBSTANCE IN OUR SEAS

OIL: CAN'T LIVE WITH IT, CAN'T LIVE WITHOUT IT

Sadly, we would be lost without oil and its byproducts. Oil provides about 40 percent of the energy Americans consume, and about 97 percent of our transportation fuels. For better *and* for worse, our society has accepted surrounding itself with everyday products whose base component is *petrochemicals*, or crude oil. From crude oil comes gasoline, home heating and diesel fuel, jet fuel, heavy oils for industry, power for marine transportation and electric power generation, still gas, coke, liquefied gases, petrochemical feedstocks, asphalt, road oil and lubricants.[1] 1994 estimates report world consumption at *66 million barrels of oil each day.* [2]

We seem unable to control our demand for crude oil. But what does this oil dependence really cost us?

THE CARIBBEAN CHOKE POINT

The Caribbean sits directly in the center of one of the world's six major trade routes, commonly referred to as *choke points* because they are subject to incredible shipping congestion. There are 73 oil refineries in the Caribbean, handling on average some 12 million barrels a day. At least nine of these refineries are in coastal areas. To further complicate matters, a pipeline for trans-shipment of Alaskan crude crosses Panama. Facilities to transfer Arabian crude from super tankers to smaller tankers are situated in Bonaire, Curacao, the Bahamas and Grand Cayman.[3] Half a million barrels of oil a day pass through the Panama Canal, a passage of about 50 miles.[4]

As long as shipping by tanker is the most effective means of moving oil from its source to energy-hungry countries around the globe, and so long as the demand for oil keeps growing, traffic along these choke points will increase. The U.S. imports most of its crude oil. In 1993, the U.S. consumed imported petroleum products in the staggering amount of 3,146,454 thousand barrels, or *8.62 million barrels a day*. This figure increased to a record high in 1994. U.S. Energy Information Administration data indicates that in one year our consumption increased by 3 percent to 8.9 million barrels a day. According to a recent report, "For the first

ILLUSTRATIVE WORLD CRUDE OIL FLOWS AND SHIPPING LANES

① **Strait of Hormuz**: Connects the Persian Gulf with the Gulf of Oman and the Arabian Sea
– 70 tankers and 14 million barrels per day (mbpd)

② **Strait of Malacca**: Connects the northern Indian Ocean and the South China Sea and the Pacific Ocean
– 41 tankers and 7 mbpd

③ **Suez Canal in Egypt**: Connects the Red Sea with the Mediterranean Sea
– 10 tankers and .9 mbpd

④ **Bosporus**: Connects the Black Sea with the Aegean Sea and the Mediterranean Sea
– 20 tankers and 1.6 mbpd

⑤ **Rotterdam Harbor**: Connects the North Sea with the Rhine River
– 8 tankers and .6 mbpd

⑥ **Panama Canal in Central America**: Connects the Pacific Ocean with the Caribbean Sea
– 5 tankers and .5 mbpd

Note: The arrows represent an approximation of the corresponding data.

(Source: *Petroleum Supply Monthly*, January 1994. Reprinted with permission from Energy Information Administration, U.S. Department of Energy.)

time in U.S. history, petroleum import supplied over 50 percent of our nation's demand" for fuel.[5]

TANKER TRAFFIC ON THE RISE

These figures add up to a disturbing forecast. In order to meet our daily needs for petroleum products, increasing numbers of tanker vessels will travel the seas along established shipping lanes, creating a *petrohighway* to rival our interstate highways. No one knows for sure how much traffic will increase, but experts agree that oil shipment across oceans is on the rise. According to one report, "It is anticipated that seaborne oil exports from the Persian Gulf will jump almost 30 percent by 1997 from 1991 levels. There will be a worldwide rise of 16 percent in the volume of seaborne crude oil trade, with a 29 percent hike in movement of refined products by tanker."[6]

U.S. oil imports are projected to increase annually at 2.5 percent through the year 2010.[7] But recall that preliminary figures suggest that *actual* consumption increased 3 percent between 1993 and 1994.

According to the Natural Resources Defense Council (NRDC), a nationally recognized organization that tracks problems related to oil transportation on the oceans, "In the Gulf of Mexico alone, tanker traffic is estimated by the state of Louisiana to double by 2007."[8] Another organization, the National Research Council (NRC), summarized the tanker spill threat with dire predictions: tanker traffic in U.S. waters is on the rise with an expected increase in the import of crude oil and petroleum products of 50 percent by the year 2000. Ships carrying this supply are 80 percent foreign-based, not U.S. flagships.[9] Remember, these figures are only predictions. Actual numbers will most likely be higher.

Much of this tanker traffic takes place by tank barges. According to a July 1994 USCG study of oil pollution events in U.S. waters between 1985 and 1991, tank barges were identified as the leading source of pollution events where fluid products were spilled. Towing vessels, directly associated with the transfer of these tank barges, have an *uninspected* status, which leaves a gap in terms of the USCG enforcing inspections and setting manning requirements.[10] As a result, vessels such as the *Morris J. Berman* can operate with poorly maintained vessels, poorly trained and undermanned crews, and in a substandard fashion. To further complicate matters, barges operate in closer quarters than large ships. Most barge voyages are coastal, close to coastal hazards (as well as important reef/ mangrove and seagrass ecosystems) and move through more crowded seaways, which require competent piloting skills.

We need more oil – but does that mean we must simply tolerate more oil spills? The risk of oil spills increases annually, in large part because the world's aging tanker fleet grows ever older. "In 1993, the average age of the 80,655 ships in the world fleet was 18 years," according to Lloyd's Register of Shipping. The fleet will age before being replaced because freight rates make it almost "impossible" to buy new ships and still operate at a profit. The fact is, older ships have more accidents.[11]

Added to the age problem are economic considerations. The crew size on tankers is shrinking: "Twenty years ago, the average cargo ship or tanker might have a crew of between 40 and 50. Today the average ship could well be operating with fewer than 20 people on board."[12]

Bottom line thinking translates into more casualties. Minimum maintenance of aging vessels seems to be the rule, not the exception. A London-based salvage association reported that nearly 50 percent of all claims in 1991 came from machinery-related casualties that were the direct result of crew negligence and lack of maintenance. A related cost-cutting factor is that fewer people performing more tasks suffer more fatigue. Of 400 officers surveyed, about 40 percent declared their work load "too taxing" or "absolutely overtaxing." Younger hands are replacing older, more experienced sea crews. These less experienced seamen seem willing to endure longer work days of 12 to 14 hours, with broken sleep and little in the way of benefits. Multinational crews are cheaper to employ. Frequently, 20 crew members may speak different languages from each other. Impaired communication complicates the problem of fewer personnel. Poorly trained crews and lack of a common language between crews, officers and land-based workers can result in inefficient ship operations. During a crisis, crew training and communication can be critical to the safety of the ship.[13]

As predicted, the end result is shipping catastrophes and oil spills.

SITTING DUCK

The tiny island of Puerto Rico, the site of the *Berman* spill, is an example of how our oil needs play out in real life. Puerto Rico is situated in the middle of many established shipping routes and at the heart of the Panama Canal choke point. Puerto Rico is the eastern most island of the Greater Antilles, bordered on the north by the Atlantic Ocean and on the south by the Caribbean Sea, and relies exclusively on maritime shipping. The island requires regular supplies of fuel oil for its energy needs. Tankers and barges regularly pass by, taking the Mona Passage between this island and the rest of the Caribbean. Ships also call at 13 commercial ports on the island: three on the north coast, two on the east, eight on the south, and one on the west. San Juan Harbor is the third-largest port in the Caribbean with 20 marine cargo terminals handling 80 percent of all cargo entering and leaving the island.[14]

This island has been a magnet for tanker and barge accidents. Since the early '60s, Puerto Rico has been on the receiving end of several major oil spills:

- The *Ocean Eagle* in March 1968 spilled some 83,400 barrels (3,502,800 gallons) of light Venezuelan crude into San Juan Harbor when the ship broke in two. Sixteen miles of the Condado Beach area were oiled.

- The *Zoe Colocotronis* in March 1973 spilled some 37,579 barrels (1,578,318 gallons) of Venezuelan crude off La Parguera, a protected reef area on the southwest coast, when the captain jettisoned the cargo into the sea to move the ship off a reef. Two and a half acres of mangroves were destroyed.

- The *Z-102* in December 1975 spilled some 7,679 barrels (322,518 gallons) of Bunker C fuel oil when the barge grounded on a reef at the northwest entrance to San Juan Harbor. This accident caused slicks from Isla Verde (the airport area) to Dorado Beach on the north coast.

- The *Peck Slip* in December 1978 spilled some 11,000 barrels (462,000 gallons) of Bunker C when the barge struck bottom near Cape San Juan. [15]

- Most recently, in January 1994, the *Morris J. Berman* grounded when the tugboat lost the barge and it ran onto a reef. Some 19,000 barrels (798,000 gallons) of No. 6 fuel oil poured into the sea at Punta Escambrón near the Condado tourist area, with the balance left in the holds of the ship likely to leak over time.

The official report states that an average of 50 oil spills occur annually in Puerto Rico.[16] Other sources say this number is far too low and that spills of oil (of all sizes) occur almost daily.[17] Certainly not all spills are of catastrophic size or impact. The typical spill may be an accidental or intentional release of only a gallon of diesel fuel by a day boater into the sea. But each input adds to the problem.

EPHEMERAL DATA

Spills from tank vessel accidents represent only a modest amount of all oil that flows into the sea. Conservatively, tanker accidents, whether tank ships, tank barges or towing vehicles pulling tank vessels, account for only 5 percent to 10 percent of all oil discharged in the ocean. The other 90 percent to 95 percent comes from land-based sources, such as improperly disposed of dirty oil from automobile engines.[18]

It is difficult to determine how much oil is spilled into the world's seas. Reports from various organizations that track this information are contradictory. Industry depends on the spiller to provide numbers in many instances, or on unrelated individuals who witness a spill. Reporting is less than accurate. Information has been described as "ephemeral data," short-lived, transitory and hard to define.[19]

The International Tanker Owners Pollution Federation Ltd. (ITOPF) provides annual spill figures. This federation of tanker owners from around the world joined together to insure losses for their members from oil spills. Since 1974, ITOPF has maintained a database of oil spills from tankers, carriers and barges, covering all accidental spillage (except for those caused by war). According to ITOPF, 138,000 tonnes (metric ton, a unit of 1,000 kilograms) of oil entered the ocean from tanker accidents involving spills of over 700 tonnes (5,131 barrels) in 1993. Data for unreported smaller spills (of less than 50 barrels) is incomplete and would increase this total.[20]

Examining the amount of compensation paid out for damage from oil spills is another way to approach the subject. In 1969, the International

Convention on Civil Liability for Oil Pollution Damage (CLC) was ratified to provide compensation for victims of spills. This was supplemented by a second treaty, the International Convention for the Establishment of an International Fund for Compensation for Oil Pollution Damage, 1971, as amended (Fund Convention). Each year, the International Oil Pollution Compensation Fund (IOPC) publishes oil spill statistics. This organization is authorized to pay out money contributed to the fund by treaty signatories who receive more than 150,000 tonnes of crude oil or heavy fuel oil carried by sea.

As of the date of the *Berman* spill, IOPC paid compensation in 63 incidents (from a total of 69) between 1978 through December 31, 1993, 35 incidents in Japan, 22 in Europe and the others throughout the rest of the world. Compensation was paid in most of those cases. According to IOPC, the Fund paid out compensation for oil spills totaling approximately $103,941,927 (66,757,821 pounds sterling) between its inception and December 31, 1993.[21] (Given a conversion rate of $1.557 U.S. dollars to 1 pound as of September 1994.)

To put IOPC figures into perspective, consider that (1) the U.S. is not a signatory to this treaty, so spills occurring in U.S. waters are excluded; and (2) IOPC will not pay any costs associated with loss of the natural resources affected by spills, such as beach property, historic structures, coral reefs, mangroves, coastline damage, seagrass beds, and all the myriad of living and structural beings contacted by or near to released oil product. The cost of restoring these marine organisms or coastal structures to pre-spill state is astronomical. The public's lost use value of these resources is not included in this amount.

An expert working in this field has charted how much oil spills into American waters (including territories and commonwealths such as Puerto Rico, the U.S. Virgin Islands and Guam) from groundings, collisions, or structural failures of barges and tankers. The information relates to larger spills of over 10,000 gallons.

According to this summary, in 1994 there were nine spills in U.S. waters of over 10,000 gallons each from tankers and tank barge collisions, groundings or structural failures, for a total discharge of some 833,400 gallons of oil, including *Berman*. Using *Berman* as an example, the OSIR figures are understated because of several drawbacks to the database:

- OSIR data is compiled from original spill amounts as reported by the press.
- Often the *total* amount spilled is greater than the *initial* amount reported by the press, but no later correction is reported.
- OSIR only includes oil spills reported by the press in its database.

NUMBER AND AMOUNT OF OIL SPILLS IN U.S. COASTAL WATERS FROM TANKERS AND TANK BARGES CAUSED BY COLLISIONS, GROUNDINGS, OR STRUCTURAL FAILURES

Year	Number of Spills	Total Amount Spilled (Gallons)
1989	16	14,371,000
1990	19	7,210,000
1991	9	187,000
1992	2	40,000
1993	5	384,500
1994	9	833,400

NOTES: • Analysis of Oil Spill Intelligence Report data.

• Includes spills from tankships and tank barges only.

• Includes oil spills of 10,000 gallons or more only.

• Includes all U.S. States and territories.

• Includes spills on coastal waters only. For purposes of this analysis, harbor areas such as approaches to New York, Houston, and New Orleans were included. Spills on the Gulf Intra-coastal Waterway were also included.

• Includes only spills resulting from accidental discharges such as groundings, collisions, and structural failures. Does not include discharges occurring at the pier.

Prepared by Gregory DeMarco, Project Manager, ICF Incorporated, Fairfax, VA.

The amount spilled in *Berman* according to U.S. Coast Guard records was actually 76,000 gallons *more* than reported by the attached table, a significant misstatement. Correcting the initial figure results in a higher amount of oil discharge as reported from this discrete source, totaling some 909,400 gallons in 1994. Given OSIR's data limitations, actual amounts of oil spilled in U.S. waters from tanker and barge casualties are likely much higher than reported.

Other significant sources of oil discharge from shipping are not included in these discharge figures, such as spills from cargo ships, freighters, recreational vessels, oil lost during transfer operations between vessels, accidental or intentional discharges from pumping of bilges directly into the sea, and small discharges occurring during routine operations at coastal oil facilities.

The American Petroleum Institute recently completed a study which indicates that the amount of oil spilled from *all sources* (including land-based facilities and other sources not included in chart figures) into U.S. waters in 1994 rose dramatically from 1,993,000 gallons in 1993 to 3,947,000 gallons in 1994, due in large part to *Berman* and a pipeline rupture accident in Highland, TX. However, even with these two catastrophes, the trend is in the right direction. OPA 90 appears to be working. Prior to its 1990 passage, the annual average oil spilled in U.S. waters from all sources was 5.7 million gallons annually.[22]

GLOBAL POLLUTION

In 1993, a blue-ribbon panel of scientists and notable experts in the oil and marine fields produced the GESAMP report for the United Nations. GESAMP is an advisory body of specialized experts nominated by the International Maritime Organization (IMO), various committees of the United Nations, and other highly regarded international sponsoring bodies to provide scientific advice on marine pollution for the United Nations. This report is the most comprehensive information available today about the composition of oil, the amount of oil lost in our oceans, and what effect petroleum and its byproducts have on marine wildlife and ecosystems unfortunate enough to make contact with this substance.

According to GESAMP, the Caribbean has been hard-hit in recent years. "On average every year, about 7 million barrels (294 million gallons) of oil are dumped into the Caribbean, about 50 percent of it from tankers and other ships in violation of IMO treaties, and significant amounts from offshore oil rigs and exploratory drilling."[23]

The Caribbean is not a lone recipient. The study found oil in every ocean of the world. GESAMP estimated as of 1990 that nearly 2.35 million tonnes of oil enters into the marine environment annually from all sources. Transportation contributes 24 percent of this number (564,000 tonnes of oil). Tanker accidents alone account for 5 percent of all oil discharged. No country or continent is immune, even the Antarctic Ocean, which was subjected to two shipping spills in 1989. The problem is global:

> *"Based on nearly 100,000 observations and measurements at the sea surface,* concentrations of dissolved/dispersed petroleum residues ... were present nearly everywhere (emphasis added) ... *In addition, data demonstrates that floating forms of oil pollution (i.e. tar lumps) are closely associated with the tanker lanes and other areas of ship activity and that the distribution on a larger scale can be accounted for in terms of transport from these areas of input by surface ocean currents."*[24]

WHAT, EXACTLY, *IS* OIL?

Petroleum is the term commonly used to describe crude oil and a range of refined oil products. Simply stated, crude oil is created by the transformation of fossil organisms (marine or plant material many millions of years old) into a liquid mixture found underground. On a molecular level, crude oil is a mixture of thousands of compounds called *hydrocarbons* and other elements, such as sulphur, oxygen, nitrogen and trace metals.

Hydrocarbons are molecules consisting of two base elements, hydrogen and carbon. Different classes of oil are distinguished from each other by the way their atoms are linked together. *Aliphatic* oil consists of straight or branched chains of carbon atoms. This highly toxic substance causes cell damage in low concentrations, and death of organisms at high concentrations. *Alicyclic* oils have carbon atoms arranged in a circle. Scientists do not yet understand enough about this class of oil to know how poisonous it is to marine organisms. *Aromatic* oil has hydrocarbons arranged in a six-carbon-ring structure. This class is considered one of the most toxic types of crude oil in the short term, but it evaporates quickly, losing much of its toxicity.[25]

Most hydrocarbons are lighter than water, float on the surface of water, and are commonly referred to by spill responders as "free product."[26] Certain components of petroleum dissolve in water. Oil's soluble parts leach out, and toxins enter the water column. Air, wind, sun and water affect oil differently, depending on the type of oil spilled. The more aromatic oils, such as No. 2 fuel oil and gasoline, evaporate quickly into the air, losing as much as 75 percent of their total volume and toxicity. In contrast, No. 6 fuel oil and other heavy crude or refined products lose only about 10 percent of their mass to evaporation, leaving behind a heavy, tar-like substance.[27]

Crude oil in the coastal environment basically separates into three major categories:

- *Volatile* oil, 20 percent to 50 percent by volume, evaporates and dissolves quickly.
- *Floating* oil, which becomes the mousse-like tar described above.
- *Sinking* oil which falls to the bottom of the ocean floor.[28]

Another way to characterize oil, particularly in the situation of spill response, is as *non-persistent* or *persistent* oil. Non-persistent oils tend to disappear quickly from the sea surface and include such oils as gasoline, kerosene and diesel fuel. Persistent oils dissipate more slowly and usually require a cleanup response. Most crudes and refined residual oils, like the oil in *Berman*, have varying degrees of persistence, depending on physical properties and the amount spilled.[29]

One of the following fates awaits oil spilled at sea:

- Oil evaporates into the atmosphere.
- Oil disperses into the water column as fine droplets.
- Oil dissolves and eventually becomes part of ocean bottom sediments.
- Sunlight oxidizes oil into carbon dioxide.

FATE OF SPILLED PETROLEUM IN AN AQUATIC ENVIRONMENT

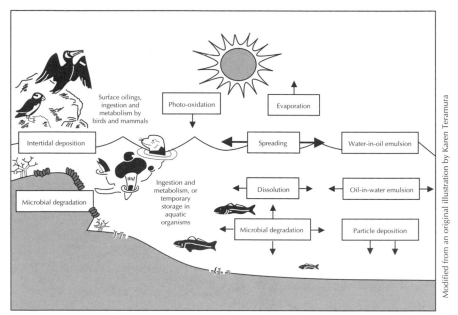

Modified from an original illustration by Karen Teramura

(Illustration used by permission from Albers, P.H. 1992. Oil Spills and Living Organisms. Publ. B-5030, Texas Agricultural Extension Service.)

- Microorganisms and other marine creatures metabolize oil. [30] (GESAMP, 1993).
- Human responders recover spilled oil.
- Tarballs, consisting of compounds (with as much as 20 percent crude oil) found covered by a skin, remain after the above processes take place and may persist for years.[31]

MOUSSE

Following an oil spill, a *sheen* develops on the surface. This oil layer of molecular thickness appears in sunlight to possess color. Sunlight causes the aromatic portion of oil to evaporate, usually within a short time after the spill.

The remaining oil enters the water column by a *weathering* process of wave action, currents and sunlight that react to make leftover oil more dense and unable to float on the surface. When you mix common household olive oil with water, the oil floats. Once you shake this mixture in a container, particles of oil become suspended in the water, creating a new substance. In the same manner, oil in the ocean either turns into an *oil-in-*

water emulsion, which disperses evenly throughout the seas, or a *water-in-oil* emulsion, which resists dispersion.

The latter compound, sometimes referred to as "chocolate mousse" because of its color and texture, forms under moderate to rough seas. This mousse emulsion is very stable and resists weathering, which is one reason why oils like that in *Berman* persist in the marine environment. Incorporation of water into the oil means that the volume of the pollutant is increased effectively three to four times. As the emulsion develops, water droplets caught in the oil become smaller and the emulsion approaches the density of the seawater. Strong turbulence can drive the emulsified oil into the water column, increasing the potential for impacting coral reefs and seagrass bed.

Mousse is one of the hardest forms of oil to recover. Ordinarily, sunlight triggers chemical reactions in oil, such as oxidation and decomposition, particularly in clear waters. Mousse resists these forces and can float and be pushed by wind and currents for long distances. Eventually, oil in every form, even mousse, is either ingested by microorganisms or sinks to the bottom to be absorbed into sediments on the ocean floor.[32]

MIGHTY MICROBES

Biological microorganisms in the ocean ingest the oil, using the hydrocarbons to produce carbon dioxide and water. There are more than 25 genera of such hydrocarbon-degrading bacteria and fungi. These living organisms attack oil once it dissolves in water, or where oil attaches to the bottom, to rocks on shore, to detritus, and elsewhere. The mighty microbes consume the oil. Fish and other marine creatures, such as bivalve mollusks and zooplankton, digest oil as well, and transform the toxic structure to less harmful compounds, including carbon dioxide or organic material.

These substances are later carried up the food chain by larger organisms, stored by predators in their cells or tissues, and released as digestive byproduct or when the organism dies. Over time, oil finally disappears in visible form from the surface of the spill site.

HOW OIL AFFECTS
THE MARINE WORLD

ULTIMATE FATE

How long does spilled oil remain in the ocean?

The answer is not an easy one. The fate and persistence of oils spilled in the marine environment depends on many factors such as the type of product, the amount, the ambient conditions at the time of the spill (tides, weather, ocean), the energy regime of the site (wave energy), the type of system (open water, shallow bays, corals, seagrasses, mangroves) and the success of the remedial response actions taken.

A number of studies have been conducted regarding the time it takes for oil to degrade to the point where it no longer affects living organisms in touch with this substance. In exposed environments, some oils can dissipate and degrade quickly (in a matter of days). Evaporation becomes the primary factor as the oil discharge spreads over an increasingly large surface. The ultimate residue of some products are tarballs which eventually wash ashore.

In sheltered areas, particularly those with fine or peaty sediments, oil which remains fluid can seep into animal burrows and into the substrate where it can persist for decades. Often oil is re-released chronically from these areas or in larger amounts during episodic events which scour or erode the substrate, causing additional damage.

FREE FLOATING

Free floating oil has the least persistency in its liquid form. Oil slicks are subject to rapid evaporation (particularly in the tropics), dispersion and degradation. Some *volatile* products, such as gasoline, kerosene and diesel fuel, evaporate within hours. Some light crudes lose up to 40 percent of their volume the first day. In contrast, heavy fuel oils barely evaporate at all. Emulsification (discussed in the prior chapter) promotes the persistence of some products. Eventually, oil degrades into very persistent tars.

As the light components of oil evaporate, a thick skin may form which protects the interior from further evaporation and degradation. These mushy tarballs release their interior fluid when stepped on. Tarballs come in all sizes. Some may be smaller than a thumbnail, while others may be more than a foot across. Tarballs eventually strand along shorelines or collect particles which increase their density and then sink. They may resuspend in the water, like those in *Berman*, with heating and wave action.[1]

Other scientists express differing opinions about the time it takes to degrade free floating oil. Dispersed in the water column, free floating oil typically remains for less than six months, unless the spill occurred right before winter at high latitude when it becomes trapped under ice until the following spring.[2]

According to Dr. William Taft, a marine biologist in Florida, certain components of heavier free floating oil, naphthalenes for example, may be toxic and persist for years and even decades. The lighter-weight fractions degrade more easily and may be gone in hours, days or weeks.[3]

TRAPPED OIL

Scientists who have studied oil spills which affected mangroves and shallow reefs have found that oil can persist for decades in these sheltered environments. They found that medium-weight hydrocarbons persisted for five years after a massive oil spill near the Galeta Marine Laboratories of the Smithsonian Tropical Research Institute in Panama (*Bahia Las Minas* spill, 1986). Some scientists predicted the continued existence of oil for up to another 20 years.

The Bahia Las Minas oil spill involved a shore-based tank rupture that spilled between 8,000 and 14,000 tonnes (50,000 barrels) of medium-weight crude oil into mangrove forests, seagrass beds and coral reefs. Mangrove roots showed no recovery six years after the spill. Hard coral recovery was estimated at decades. Oil in toxic quantities remained in the deep seagrass beds, those not totally destroyed by the initial oiling, for five years after the spill. The mud surrounding the mangroves threatened to be a longtime reservoir for continuing and chronic re-oiling of the coastal systems.[4]

Other studies confirm the long-term existence of oil trapped in sediments, including a heavy spill in West Falmouth, England in 1969, a tanker grounding in Puerto Rico, and the huge oil slick resulting from the Gulf War in 1991. In the first two instances, oil hydrocarbons were still detectable in sediment 20 years later.[5]

TENACIOUS, COHESIVE AND PERSISTENT

The *Morris J. Berman* spilled No. 6 Bunker C fuel oil, a substance produced from the residues of refining oil consisting of a mix of other oils diluted with other chemical compounds. This heavy home heating oil weighs almost as much as water. It has been described as "tenacious" and "cohesive," meaning it floats or remains almost intact. Only a fraction dissolves in water, and only a small portion, the aromatic component, is highly volatile and evaporates quickly.[6] What remains sinks to the ocean floor to lie on top of the sand and bottom organisms, and essentially creates an *asphalt highway*. When high wave action stirs the water or high temperatures heat the air and water surfaces, the oil resuspends as chocolate-size malt balls or floating tar strands.

HARD ON REEFS

Many coral reef experts express their concerns about the harmful effects of oil spills. However, because so many factors affect the exposure of oil to corals, it is not surprising that scientists have not reached agreement on how spilled oils affect corals. Different products have different toxicities, and environmental conditions vary so that the degree of exposure changes within a given site daily.

Since the bulk of the coral reef structure is submerged, most corals are not vulnerable to direct physical coating by the oil. This is not true for corals in the upper part of the reef (the reef crest) which are exposed to the atmosphere during low tides. This exposure is longest and much more extensive during spring tides. Oil stranding on an exposed reef crest can have disastrous consequences, because of its smothering effect, toxicity, and because the black oil rises as it warms under the tropical sun. Oil becomes trapped in cavities, crevices and fissures in the reef crest, acting as an *oil reservoir* which leaches out for long periods of time, well after the spill.

Subsurface corals are vulnerable to the water-soluble toxic fractions of the spilled product. Turbulence created by breaking waves at the reef breaker zone can drive globules into the water column and expose the coral polyps within the seaward reef slope to the dispersed oil.

The sheltered shallow water habitats behind the reef crest are also sensitive to oiling. Water moves sluggishly over very shallow areas, increasing the risk of direct contact, toxicity and extreme warming. Numerous sea urchin and invertebrates may be killed by oil leaching from the reef crest into the sheltered lagoon habitat.

As a result of all these variables (toxicity of oil, amount of reef exposure), oil can be expected to have a wide spectrum of impact on coral reefs and their associated biota, from short-term, limited impacts to massive mortalities.[7] In the *Berman* spill, biologists, chemists, oceanographers and specialists involved expressed their opinion, with minor exception, that oil harms the sea and its marine organisms:

Antonio Mignucci-Giannoni, Scientific Coordinator for the Wildlife Mortality Assessment in *Berman*: *"What is the effect of oil on hard corals? Well, what you end up with is* a dead reef (emphasis added)."[8]

Jerry O'Neal, USFWS scientist: *"There's a filter effect. Oil on the surface can 'eliminate birds off the face of the earth,' kill the phytoplankton, cause fish to*

141

go belly-up and hard corals to die. The oil continues to filter down to the deep ocean bottom, where it affects the benthic organisms as the hydrocarbons settle into the sediments. The larval young die and the adults stop reproducing. Deep sea currents resuspend the oil and carry it far away from the initial spill, affecting sea creatures elsewhere. Tarballs coat the bottom sand, taking years to degrade and harming the microscopic organisms living in the sand. These plants and animals form the base of the food chain, so their absorbed toxicity filters up to us eventually. Oil is like an herbicide." [9]

Carlos Padin, planner in the Puerto Rican DNR: *"Oil in the Condado Lagoon after the* Morris J. Berman *spill so significantly harmed the seagrass beds, and even worms and other creatures living two feet under the sand, that the whole place looks like a desert."* [10]

James Timber, scientist with the Puerto Rican DNR: *"Oil is made up of highly volatile and toxic components. Whatever substrate on a reef comes into contact with the oil at first will suffocate and die. An oil spill is like spraying insecticide on the corals (emphasis added)."* [11]

Gilberto Cintron-Molero, biologist and international affairs specialist with USFWS: *"Many types of oil are very toxic and act like poison, killing reef animals directly. If the tide goes down, the oil physically suffocates the reef. Dispersant used to break down the oil in itself may be very toxic to shallow water reefs. Eventually oil is broken up into small particles and sinks. If it's not ingested by microbes, the oil forms tarballs and coats the ocean floor."* [12]

Aileen Velazco-Dominguez, chief scientific coordinator for the *Berman* spill for Puerto Rico's DNR: *"Reefs suffocate. Coral is like the bones in the body of a person. When the hard corals go, the rest of the reef follows."* [13]

Vance Vicente, biologist and oceanographer consulting for NOAA during *Berman*: *"What's the physical impact of oil? That depends on the type (crude is not as toxic as refined), the amount spilled, the season and the depth of the spill. Generally the sandy beach communities where the crabs, crustaceans and phytoplankton live are very vulnerable. This community is the food source for fish. The intertidal rocky shoreline members (when the surf goes up into the higher part of the rocks) are also very vulnerable, such as the shells, whelks, octopi, fish, algae and sponges. Submerged marine communities are not usually hit, except if the oil stays in the area after the aromatics evaporate. Then the thick oil sinks and covers the corals, gorgonians and seagrasses to their detriment. Seagrasses are the nurseries for reef fish, lobsters and other animals which serve as the diet of the manatees, green turtles, herbivorous fish and sea shells. Oil in the sediment is stirred up and ingested by the juvenile fish and in turn by these larger sea animals."* [14]

BLACK BACTERIAL SLIME

Coral can defend itself against spilled oil – to a limited extent. Oil affects coral in several ways:

- Directly, if a reef is exposed, such as low tide when the top of the coral is exposed above the water line.
- Indirectly, when oil is distributed in the water column through dispersion when evaporation and sunlight cause weathering and oil sinks.
- When oil combines with other materials in the water, such as mineral particles and debris, and sinks onto the reef.
- When oil sinks onto bottom sediment and is then resuspended due to wave and current action.[15]

However oil reaches coral, the polyp's reaction is automatic. The coral retracts into its calcium shell and secretes a protective layer of mucus. Oil droplets stick to this tissue. To some extent, the polyp can later shed this layer and rid itself of oil. When the oil coating remains too long or is in too great a quantity, the coral polyp and other reef creatures smother to death. Ultimately, the stress caused to coral reefs may result in bleaching, tissue swelling, mucus overproduction, impaired reproductive functioning, and dead areas of a formerly healthy reef. Mucus produced by the coral contains oil, which is ingested and brought into the food chain by mucus eating fish and crustaceans.[16]

Marine scientist Eugene Kaplan graphically described the resulting imbalance caused by oil spilled on reefs. Even when coral survives direct or indirect impact, the increased mucus production attracts larger numbers of bacteria than are normally found on the reef. These organisms engulf the coral and eat everything down to the reef skeleton. According to Kaplan, a once live coral reef becomes a field of "black bacterial slime."[17]

DIFFERENCES OF OPINION
The following is a review of various studies in which scientists discuss the effects of oil on coral, and how long it takes coral reefs and their fringe communities to recover after a spill.

Lab Studies: Controlled Events
In his research, biologist Eugene Shinn found that brain corals exposed to controlled amounts of oil for minimum time yielded quick recovery within 48 hours after the oil was removed.[18]

Another marine scientist, Phillip Dustan, studied the impact of exploratory and shallow water oil wells on several different sites in the Florida Keys. Drilling took place between two and 24 months prior to the study. The wells were one-time rigs with no blowouts, subsequent spills or slow leaks creating chronic oiling. Dustan's team found no permanent damage at any site surveyed, except one in which the legs and anchors of the drilling rig abraded the reef. All other sites displayed strong coral growth of the original coral and/or recolonization by new coral.[19]

A well-regarded reef expert, Dr. Richard E. Dodge, has conducted numerous experiments on the effect of oiling on coral. In one controlled lab study, his team exposed coral in tanks over a four-week period to oil and a mix of oil and *dispersant* (chemicals used to break up oil and disperse it into the water column in a manner not unlike using soap to wash your hands.) Dodge equated this study to conditions that might exist in an actual spill in shallow water. Once the team removed the oiled water from the tanks and circulated clean water in them, the corals showed recovery between two hours and four days, except for those polyps whose tissue burst and became infected with bacteria.[20]

In yet other spill experiments, Dodge and his coworkers found that more chronic, long-term oiling conditions produced very negative results: permanent damage to the coral polyp's mouth opening, persistent disruption of normal activities such as feeding and defense mechanisms, pigmentation loss and rejection of zooxanthellae, resulting in death.[21]

In another controlled study involving 2.5 years of field experiments with oil in crude or crude-and-dispersant mixed form among coral reefs, mangroves and seagrass beds on Panama's Atlantic Coast, Dodge and his associates attempted to reconstruct what they believed would be a typical oil spill "worst case" scenario. They spread oil or oil and dispersant mix over different sites for a period of 24 hours, then collected as much of the remaining oil as possible. The results showed minor effects on seagrasses and corals, but major effects on mangroves, because oil became trapped in the sediment and continued to re-release even 20 months later. They predicted recovery of coral so oiled to begin within one year to 20 months, but expected the mangrove forests to take 10 to 20 years to recover.[22]

Field Studies: Actual Spills

Marine biologists at the Smithsonian Tropical Research Institute in Panama believe these controlled experiments do not equate to real-life oil spills. Biologist Hector Guzman said, "Contrary claims that oil has little harmful effect on corals (e.g. Shinn 1972, 1989) are based on experiments, the results of which appear not to scale up to the effects of a major oil spill."[23] Studies conducted of the Bahia Las Minas oil in Panama provide a telling picture. Of the 50,000 *barrels* spilled in Bahia Las Minas in 1986, reportedly only 60,000 *gallons* were recovered, leaving the balance to spread out over an 85-kilometer stretch of coastline.

One major factor that distinguishes the Bahia Las Minas studies from others is that a Smithsonian Tropical Research Institute is located close to the site, so that considerable *baseline data* of the area exists prior to the spill. Baseline data is research over time that provides critical measurements of the natural fluctuations of a particular ecosystem. A second important difference is that the Bahia Las Minas studies were conducted over approximately five years, allowing scientists a much longer observation period to study long-term effects of oil.

The results from the Bahia Las Minas studies were startling. Researchers found high mortality along the seaward side of the fringing reefs. Two years after the spill, according to one study, "Numbers of coral, total coral cover, and species diversity based on cover decreased significantly with increased amounts of oiling."[24]

Oil continued to reemerge from sediment in the mangrove forest. In certain sites in the soft soils around the mangroves, oil remained in a non-degraded state years after the spill. Researchers described this intact oil as "a time bomb being released slowly."[25] While there was extensive death of reef corals, seagrass beds and mangroves from the initial spill impact, long-term impacts were even more profound. Study results concluded:

- Most subtidal seagrass communities survived intact.
- Seagrass communities were damaged severely in the intertidal zone (area between high and low tide lines).[26]
- Mangroves were affected down to their roots, with a loss of as much as 74 percent vegetation in the fringe habitat.[27]
- Marine creatures living in the mangrove root systems such as sponges, anemones, corals and hydroids were almost wiped out for up to five years when some, but not all, reappeared.[28]
- Serious shoreline erosion resulted as a side effect of the loss of natural barrier mangrove, seagrass and coral habitat.[29]

Researchers concluded, *"This summary is intended to alert the environmental community that toxic effects of hydrocarbons will probably persist for at least 20 years in deep mud tropical coastal habitats affected by catastrophic oil spills."* [30]

Keller and Jackson answered the question, "Does oil affect reef corals?" with an emphatic yes. They report these impacts from the long-term oiling in Bahia Las Minas:

- A reduction of live coral cover, abundance and diversity.
- Correlation between injured corals and the extent of oiling.
- Decrease in coral growth rate.
- Decline in fecundity of reef-building coral.
- Little prospect for rapid reef recovery.

The neighboring mangroves and seagrass beds fared just as poorly. Mangrove forest recovery rates were stunted by 50 percent. In some cases of heavy oiling, seagrass community beds died out totally. Certain types of echinoderms found in those beds were completely absent six years after the spill. The average time for repopulation of oiled mangroves and seagrass areas to equivalent prespill levels was estimated at 50 years. Scientists concluded: "Oil trapped in these environments can persist and remain toxic for decades."

A BAD MIX

Oil is a poisonous substance that affects not only coral reefs and their neighboring ecosystems, but also every type of marine organism it touches, including *invertebrates* (creatures without a backbone or segmented spinal column) and fish, birds, mammals, plankton and plant life. The harmful effects can best be generalized for all species:

- Larvae and juveniles are particularly susceptible, and often die on direct contact or when they ingest oil.
- Adults are affected when touched directly by oil, when they eat oil or oiled materials, or when, as larger animals in the food web, they eat smaller organisms who have ingested petroleum. The kill rate of juveniles or smaller prey leads to a loss of food source.
- Death occurs when oil coats skin, feathers and gills, and animals are asphyxiated.
- Species die indirectly when hypothermia sets in due to exposure. Oil destroys the animal's natural ability to ward off cold.
- Marine organisms far removed from the spill may experience stress from oil invasion of their neighboring environment, and from sublethal ingestion of oil. Reduced resistance makes them more susceptible to infection.
- Marine plants and wildlife may experience changed habitat, with a resulting significant disruption of usual functions such as feeding and reproduction.[32]

Consider the effect of oil on a sea bird. One drop of crude oil can kill the unborn bird inside its egg. More mature birds are poisoned when they drink oiled water or consume prey who have been in contact with oil. Birds spread the oil over their bodies, either through direct contact or preening of their feathers. The oil affects the normal functioning of the bird's feather structure. The oiled bird loses its ability to retain heat, to fly, to hunt and to feed, resulting in death by hypothermia, starvation, or inability to elude predators. Even when an oiled bird does not die immediately, there are usually *sublethal* (short of death) effects, such as disease, retarded growth of young, and improper organ function. Anomalies form in gonadal tissue and spread to other organs or enter the bloodstream in a form like leukemia. Ulceration of the intestinal lining inhibits digestion. Pneumonia can set in.

Overall environmental impacts are significant. Food sources are depleted through contamination. Nesting habitat is reduced, causing less successful reproduction. Future generations are affected as the natural reproductive cycle is disrupted.[33] Mammals that depend on their fur for insulation are equally vulnerable. The *Exxon Valdez* spill generated indelibly etched images of significant sea otter kill-offs, as the animal's fur became so matted with oil that it could no longer trap air or dispel water. These mammals died directly from oil digestion, or indirectly from hypothermia. Sea turtles have difficulty distinguishing tarballs from familiar food and eat the oil, or petroleum comes in contact with turtle eggs and newborn hatchlings, usually with deadly results.[34]

Fish and other marine life depend on chemical communication for feeding, selection of habitat, fleeing predators and sexual attraction. Oil pollution interrupts or stops these basic functions from taking place.[35]

Even the tiniest creatures succumb to oil. In experiments with marine bottom algae exposed to crude oil, these organisms developed negative reactions and died within five to 15 days.[36]

Ingestion of oil by simpler life forms spreads upwards. A study of crayfish subjected to No. 2 fuel oil revealed oil accumulations in the liver, gall bladder, fat, kidneys and blood of the ducks who consumed them.[37]

William Taft summarized the effects of the West Falmouth spill in 1969 when crude spilled into the mouth of a semi-enclosed harbor and neighboring marshlands. His message: *Petroleum and sea creatures don't mix.*

> *"The number of marine organisms declined from 200,000 per square meter to 2 animals/square meter after the first week. After the second week, no more deaths occurred – there were none left to die. All scallops died. Four years later, after recolonization, shellfish were still too oil-tainted to be eaten. Oil permeated the peat at the bottom of the marsh, and after four years persisted in the sediments to depths of five feet."* [38]

The volatile component of oil affects surface organisms like plankton. The floating or mousse component suffocates shoreline invertebrates, and reduces the insulation ability of bird feathers and mammal fur. Oil's sinking component strikes mud and bottom-dwelling communities. Even when the initial effects of oil do not cause immediate death, many studies indicate that long-term impacts are far-reaching in what is described as a *trophic cascading* effect: what happens to one species affects all the others who depend on and are part of its community web. Within a species, there is strong support for findings of major habitat disruption in population, behavior, reproduction and feeding ability for periods of two to 15 years, depending on the type of community, oil composition and other factors.[39]

IMPERFECT WORLD

Almost every American heard or read about the *Exxon Valdez* spill in 1988. This spill affected miles of virgin Alaskan coastline and brought destruction to thousands of creatures in contact with the 11 million gallons of petroleum that invaded Prince William Sound and spread outward. According to Rick Dawson, a marine biologist and member of the USFWS team brought in to fight the spill:

> *"There were only two other spills worse than the* Exxon Valdez *in the world. We spent between a quarter and a half million dollars trying to assess the damages of this catastrophe. We were still in the process of trying to determine the damages when Exxon settled. One of the problems we faced was just how many resources can you throw at recovery of marine creatures hit by the oil. For example, do you spend $85,000 an otter to save maybe 30 percent of the 18,000 damaged otters, or use that money somewhere else with maybe a better overall impact? With oil spills, all you can do is your best. Those of us working on the spill are just people in an imperfect world trying to make perfect decisions."* [40]

Years later, studies reveal the extensive effects of the *Exxon Valdez* oil spill. Among the significant findings from a federal and state-issued report from April 1992:

- 3,500 to 5,550 sea otters died, with mortality continuing.
- 30,000 or more dead birds were recovered, with estimates that 10 to 20 times this mortality figure is more accurate.
- 300,000 adult murre seabirds were killed outright, with a reproductive failure equal to that number in subsequent years.
- Pink salmon experienced a 40 percent to 50 percent egg mortality in 1991, with continuing exposure of eggs to oil.[41]

The human side of the catastrophe was even worse. Exxon is now appealing a hefty jury verdict of billions of dollars for damage to man and resources.

The science of determining the effects of oil in the ocean is still in its infancy. There is no definitive answer to the question: "How long will petroleum affect this generation and future generations of animals and plants?" The marine community says it needs more hard science and careful research. The GESAMP report urges responsible scientists to establish a system of monitoring stations, to use these fixed sites for careful recovery of baseline information and then, armed with this background research, to develop effective measurements to distinguish between damage caused by pollutants and harm resulting from natural sources.[42]

Evelyn Wilcox, a highly-regarded marine environmentalist and a noted scholar on the subject of man's impact on coastal areas, addresses the call for yet more study:

> "Oil in the water column that doesn't dissipate affects animals at different stages from the larvae on up. Oil can wipe out whole fish populations. The problem is we don't know yet how serious the impact of oil is. What we know is that we are losing reefs each year due to this and other types of marine pollution, climate change, direct damage by recreation, and other human causes." [43]

What is clear is that we are facing an oil crisis in our oceans of potentially catastrophic proportions. Further study can provide useful data to support national and international efforts, including regulation and legislation, for better oil spill prevention and response.

HOW CORAL REEFS BENEFIT MAN

NATURAL BUFFERS

Reefs absorb ocean waves, preserving shorelines against erosion. Many islands were formed by coral reefs. Today these structures act as protective barriers, as *breakwaters*, for the windward side of these islands. Coral reefs create the beautiful beaches so many of us love, as the calmer waters allow sand to accumulate and certain reef organisms die, their bodies contributing to beach particles. Coral is responsible for many safe harbors. In all but the worst storms, these superstructures act like the defensive front line of a football team, keeping the powers of nature from tackling shore communities. As long as the coral keeps up with rising sea levels, these important systems will continue to buffer the land from the worst effects of the sea.

Reefs provide habitat protection for mangroves and seagrass beds, sources of nurseries for the fish and other food species we eat. These structures provide an estimated 10 to 12 percent of the finfish and shellfish harvested in tropical countries and 20 to 25 percent of the commercial fish catch of developing countries.[1] Carlos Goenega stresses the importance of coral reefs, using Caribbean reefs as his example:

> *"Fisheries in the Caribbean can be defined, with few, although significant exceptions (e.g., upwelling zones and shrimp fisheries) as coral reef fisheries ... 59 percent of the total fish consumed in Puerto Rico and the Virgin Islands come from coral reefs."* [2]

THE TOURISM ECONOMY

The reef ecosystem is the mainstay of many of the economies of coastal people everywhere in the tropical world. In the Caribbean, tourism on most of the islands is almost exclusively based on the recreational and aesthetic values of reefs closest to each island. In the U.S. in 1990, the five most crowded reefs drew more than 3,000 people a day to the upper Florida Keys. The spin-off benefits restaurants, bars, hotels, boat operators and dive shops. Store owners sell supplies to contractors, workmen, visitors and residents alike. Cabs, car rental agencies and airline operators all benefit from transporting people to the islands. Hotels, restaurants and shops cater to visitors and employ many local residents. Doctors, pharmacists, electricians, utility companies, owners of rental homes and workmen benefit directly or indirectly from the tourism industry. The list of local people in the human fraternity who enjoy revenue produced by their proximity to the reef tourist trade includes nearly everyone.[3]

NATURAL LABORATORIES

Coral reefs are an essential part of the mangroves, seagrass meadows and other complicated habitats that are home to thousands of species, including certain rare and endangered plants and animals. Next to the rainforests, these communities are among the most diverse of any on the earth. Scientists view these rare universes as important settings for the study of ecological theories and the testing of how species relate to each other and to their environment. Through global monitoring, reef systems worldwide are used to study the health of the earth.[4]

This is only one aspect of the importance of these structures and their incredible *biological diversity*. A single reef may contain thousands of fish and smaller marine organisms.[5] Each of these creatures and the reef as a whole play an important role in the web of life that includes humans. The *Gaia Hypothesis* is a fanciful name for a simple theory. According to this philosophy, the earth is a living, breathing, interconnected, self-regulating organism. The rocks, air, living organisms and ocean are all part of one symbiotic entity. The earth's environments have evolved hand-in-hand with the living creatures who inhabit the air, water and land. Loss of any one ecosystem, riverine, wetland, forest or reef affects all others. Specifically, without the biological diversity of the world's coral reefs, the remaining ecosystems will slip dangerously out of balance.

Oceanographers describe the end result as follows:

> *"In the face of environmental change, the loss of genetic diversity weakens a population's ability to adapt; the loss of species diversity weakens a community's ability to adapt; the loss of functional diversity weakens an ecosystem's ability to adapt; and the loss of ecological diversity weakens the whole biosphere's ability to adapt. Because biological and physical processes are interactive, losses of biological diversity may also precipitate further environmental change.* This progressively destructive routine results in impoverished biological systems, which are susceptible to collapse when faced with further environmental changes (emphasis added)."[6]

Why we must care about the fate of coral reefs may perhaps best be explained by USFWS marine scientist Gilberto Cintron-Molero:

"Because of species richness, a coral reef is one of the most diverse systems, full of thousands of species. When that many species live in one place, the system works. Sunlight drives the whole process. The reef and its inhabitants make use of all nutrients and light in the water. It is a very efficient and complex system. Take the reef away and you lose this wealth.

*"*Links: *There are many linkages between species on the reef and the neighboring ecosystems. What affects corals, affects other communities connected with the reef, the mangroves, lagoons and coastal shores.*

*"*Medicine: *By destroying the reefs, we may lose the prospect for new chemicals found in the sponges, soft corals and other reef animals. These may serve as a cure for cancer and other diseases attacking man. These chemicals are biologically active and very valuable."*[7]

Without its complex world of interactive marine creatures, the reef system cannot exist. When some parts of the reef system collapse, such as when sea urchins die out and alga takes over, the superstructure loses it biodiversity and thus its ability to adapt. When the reef goes, the mangroves and seagrass beds are affected. When they die off or are removed, the coastlines lose this critical barrier. As coastal areas erode, major areas of man's habitat are lost. Ecosystems are not as individual or as independent as they may seem. The collapse of one sends others nearby toppling over in a domino effect. At some point, biological ecosystems as we know them may disappear.

BIOTECHNOLOGY AND MEDICINES FROM THE SEA

Let us assume you do not care about how coral reefs affect the air we breathe and the waters that surround us. Perhaps your idea of a vacation has nothing to do with beaches and reefs, and traveling to the world's coral reefs is beyond your budget. Maybe the concept of biological diversity conflicts with your personal philosophy. There are many reasons for saying "So what?" about reef preservation.

Even if any or all of these reasons apply, you should still want to preserve coral reefs. In today's world, there is hardly anyone who has not experienced sorrow at the loss of a healthy friend or family member due to cancer, leukemia, physical deformities, mental illness and other deadly diseases or conditions. Coral reefs, like tropical rainforests, may contain natural medicines that are our best future hope in the war against disease.

Four top scientists in Puerto Rico are involved in the complex study of biomedical research. Within their specialized fields of research, they are each studying reef creatures as part of their search for cures to illnesses. One studies sea urchin embryos, trying to understand how development is controlled at the molecular level. Another studies reef animals to learn from observation new ways to modify abnormal behavior in man. Two other scientists research the toxins marine animals produce for self-defense against natural predators to help develop drugs that can defeat predatory invaders in our own bodies.

Perhaps the best way to observe their research is by sharing our conversations.

THE CASCADE EFFECT

Dr. Carlos Santiago is a molecular biologist with the University of Puerto Rico. He received his Ph.D. in molecular cell development from Ohio State and his post doctorate degree in molecular biology from Florida State University in Tallahassee. He has been working in San Juan for the past five years. Dr. Santiago received a grant for a three-year project in molecular biology.

BARBARA E. ORNITZ: Dr. Santiago, are you using any sea creatures in your research?

DR. CARLOS SANTIAGO: I work with the sea urchin and use it as a model to study early development.

BEO: What is so useful about the sea urchin?

CS: We can study the developmental events in the sea urchin from its early stage as an egg before fertilization, through each stage up to adulthood. Sea urchin eggs and the spermatocytes which fertilize them have easily recognized embryonic stages. Unlike human research, where we certainly cannot use the human embryo (the baby in the mother's stomach), with sea urchins we can obtain millions of embryos for our study. One urchin might produce millions of embryos. Human fertilization is tricky. Not so in sea urchins where we can fertilize thousands at one time.

This allows us to have a starting pool of eggs from populations fertilized simultaneously. To achieve the same result in the female population would be impossible. We'd have to find millions of women impregnated at exactly the same time, and then be allowed to use their embryos in our research. Clearly we can't replicate our studies using humans.

BEO: How do you conduct your experiments?

CS: We take millions of eggs before they are fertilized from the sea urchin. We fertilize the egg with spermatocytes and culture them in sea water. Once the egg is fertilized, it develops through the embryonic stages (morula, blastula, gastrula) *into a larval stage, and eventually into adulthood. We study these developmental stages at the molecular level.*

BEO: What do you mean by "studying development at the molecular level?"

CS: We are looking for proteins which turn on and turn off specific genes in a particular cell so that the cells can differentiate or develop into different organs or tissues. These proteins are called transcription factors. Transcription is a process of going from DNA to RNA. The RNA in turn is translated into protein, which is finally assembled into a definite type of tissue. Transcription factors turn on a different subset of genes in different cells until the genes required to make the proteins will be assembled into the type of tissue that they are supposed to become and are all turned on. In this process, we propose that there are a few master genes, nuclear genes that cause certain other genes to turn on or off, so that, for example, a liver cell becomes committed to being liver tissue and not lung tissue. We believe this cascade of events turns on immediately following fertilization and permits the egg to develop in a normal fashion. What we are looking for is what factors initiate this cascade and activate the developmental program.

BEO: How do you find these master genes?

CS: What we do is isolate the starting material from the sea urchin embryo. We extract the nuclei, purify different transcription factors, and try to understand how they interact with DNA to turn a gene on or off. It is very detailed and time-consuming work. We're trying to define the regulatory sequence by defining the protein that regulates and binds to that sequence. Once we know what factors turn on a gene, we can find that gene. We take that gene and see what turns that gene on, essentially working backwards until we find the master gene that turns on the whole process. One gene makes protein (a transcription factor) that binds to other genes, turning on first tens and then hundreds of genes in a "waterfall" effect. This is the developmental program.

BEO: How does finding this "master gene" help humans?

CS: We are dealing with basic developmental phenomena here. Only by understanding how normal development works can we hope to understand abnormal development. We need a baseline to measure the healthy gene. Then we can see how to control the unhealthy gene. What happens in cells that cause cancer is a good example of this approach. Only by understanding how cells are normally "braked" so that they don't divide wildly can we achieve a better understanding of cancer.

BEO: Do you regulate how many sea urchins you take?

CS: Yes. We use about 100 a week and take them from the lagoon near the Condado in San Juan. One sea urchin produces millions of eggs. We switch from area to area so we don't devastate any one population. Also, we fertilize some of the urchins in the lagoon to keep the population strong.

THE COMMAND NEURONS OF *BURSATELLA*

Dr. Mark Miller is a neurobiologist with the Institute of Neurobiology at the University of Puerto Rico in San Juan. He holds a graduate degree in neurobiology from the University of Connecticut and has had post-doctorate assignments in neurobiology, pharmacology and molecular biology from the University of Hawaii, Hebrew University in Jerusalem, UCLA in California and Columbia University in New York. Dr. Miller's work is similar to Dr. Santiago's, but it is geared toward an entirely different purpose.

BEO: Dr. Miller, are you using sea creatures in your research?

MM: I'm studying an animal called the Bursatella leachii *or marine mollusk. This animal is like a fuzzy snail. It is greenish-brown and can grow up to six inches in length.*

BEO: Why are you using these animals?

MM: These invertebrates (meaning spineless animals) are different from us in that they have very simple nervous systems. They have around 20,000 neurons in their systems. A human system has millions. And their neurons are very large. You can look at the brain of a Bursatella *and identify the neurons. Because they have such a very well defined behavior system, what we call a "model system," you can tell just which neurons cause a certain type of behavior. In our study of* Bursatella *we found that they tend to eat, copulate, socialize and move about in a fixed daily rhythm. For example, prior to sundown, nearly all the animals keep their distance from each other in isolated spots in the shallow water. At night they aggregate together, with as many as 88 percent of the* Bursatella *we saw touching at least one other individual. Their sexual patterns are equally as clear.*

BEO: That raises two questions. First, can't you identify which neurons in human brains cause certain behavior?

MM: I wish we could. The problem is with vertebrates like man the neurons are too small to individualize, with some limited exceptions.

BEO: Secondly, if these animals are so different from us in their neuron structure, how can they be useful for human research?

MM: Well, they aren't that different from us. Bursatella *have similar underlying behaviors and their neurons function just like ours do. Let me give you an example. Take the simple act of moving your fingers. Here's how that happens. The motor neuron releases a neurotransmitter that causes the muscles of your fingers to contract. The fingers move. In invertebrates like the* Bursatella *species, the neural circuits that are used for signalling movement function essentially the same way as they do in human muscles. What differs between us and these simple creatures is that in people any one behavior might involve hundreds of thousands of cells. In a Bursatella we can pinpoint fewer than 100 cells, for example, that organize a simple behavior. We call the approach I've just described the "identified neuron approach". We're trying to find the specific brain cells that are hooked up to other cells that cause a person to do one thing and not another.*

BEO: So they act like us and their neurons work like ours. What study are you doing with these mollusks?

MM: My job is to see how behavior is organized. What, for example, allows me to execute the three behaviors I'm doing right now, speaking to you, thinking about your questions, and writing notes to myself about what I've told you. The question I'm asking is why am I doing these three things instead of jogging, walking or water skiing. What tells my body to do one set of behaviors and not another. Something in the nervous system tells us what behaviors are compatible or not compatible with each other.

BEO: How does studying the *Bursatella* help you understand human behavior?

MM: These sea organisms spend a lot of their time copulating. What is interesting is that each Bursatella *is capable of taking the role of a male or a female animal. When it is acting as a male, the mollusk can do only one thing at a time, for example either eat or copulate. When it is being female, the* Bursatella *can undertake two activities at once, for example copulating and eating. What this indicates is that this sea creature has something inside its nervous system telling it what behaviors it can and cannot do simultaneously. We call such higher-order elements "command" systems.*

BEO: How does a command system function?

MM: We believe that certain behaviors in this animal are organized using a specified circuit configuration that may utilize principles common to all animals, including man. A specific command system triggers sexual behavior. Depending upon which is triggered, the male or the female sex, that will determine whether the animal can feed at the same it copulates. If the animal is a male, the command system may inhibit feeding behavior. If the Bursatella *is a female, the command system may not inhibit eating. Like a light switch, feeding is a behavior that is switched on or not depending upon the* Bursatella's *normal circuitry.*

BEO: How does understanding the neural organization of behavior in this model system help humans?

MM: The marine animal is a biomedical model. By understanding what circuitry causes normal behavior patterns in this creature and seeing just how that behavior system works, we can gain a better understanding of what may go wrong, creating abnormal behavior in this system. Such malfunctions may ultimately bear some relationship to the neural dysfunctions underlying behavioral and model disorders.

BEO: How does your research lead to a better understanding of mental disorders?

MM: As an example, a leading model for the cause of one mental disorder, schizophrenia, is what we call the dopamine hypothesis of schizophrenia. *What this fancy name refers to is a system in the brain. A certain molecule called dopamine is released by one neuron and affects the neighboring neurons. Dopamine is a transmitter produced in discrete regions of the brain, in cells that project widely, ie., to many areas. Such an organization is reminiscent of the command systems of invertebrates. Malfunctioning of this system is thought to underlie certain symptoms associated with schizophrenia.*

FINDING PIECES OF THE PUZZLE

Professor Abimael D. Rodríguez is a quiet-spoken man who teaches organic chemistry at the University of Puerto Rico. Born in Puerto Rico, he attended the Johns Hopkins University and received his Ph.D. in organic chemistry, followed by a two-year postdoctoral at MIT working on the synthesis of natural products and additional research at the University of Hawaii. Dr. Rodríguez is dedicated to the preservation of natural resources, which he regards as our hope for the future.

BEO: Which sea creatures are you using in your research?

AR: We work with gorgonian octocorals, what you might know as "sea fans and sea whips," and with another marine organism, the "sponge."

BEO: Those are beautiful creatures. What do you do with them?

AR: We collect the animal, bring it to our lab, and then extract certain organic components from it. These organisms, like certain other sea creatures, are unique in that they have a defense mechanism to protect themselves from predators. The gorgonian excretes mucus to stop other animals from eating it. We don't know why the sea fan makes these compounds, we're just glad to use the mucus in our cancer research.

BEO: Once you have the compound, what do you do with it?

AR: We purify the substance and establish its molecular structure. We use spectroscopy on the purified chemical, exposing the molecule to different lights and radiations. Some molecules can be subjected to X-ray crystallography, if the compound is in a crystalline form. Most of our compounds are in semi-solid form.

BEO: Why do you try to determine the molecular structure?

AR: In order to understand the action of a drug that might deter cancer, for example, you need to know its chemical structure before the pharmacologist can work with the drug. Establishing the chemical structure is like putting together the pieces of a jigsaw puzzle. Each technique we expose the compound to gives us a piece of the puzzle. We build up all the data and eventually we have the structure of the compound.

BEO: How does that work?

AR: For example, once we think we have identified the structure, we have to prove it by seeing if we can change the compound into another substance which we already know. So, we have an unknown molecule A. We do something to it, for instance, we expose the substance to certain chemical reagents. If the molecule A changes into a known molecule B, then we can verify the formula.

BEO: What is the next step of your process after you have established the molecular structure?

AR: We send the substance and molecular structure on to our colleagues for further research to see if this particular molecule can be used as an anti-cancer or anti-bacterial drug. Our work is only the beginning stage in a very long process.

BEO: Why do you use the gorgonian corals in your research?

AR: Gorgonian metabdiles possess novel structures that are largely unknown from terrestrial sources. These animals are now recognized to produce acetogenins, sesquiterpenoids, diterpenoids, prostanoids, and, in some cases, highly functionalized steroids. The natural products isolated from these marine invertebrates constitute a rich source of biologically active compounds of medicinal importance.

BEO: What do you hope to achieve?

AR: We're looking for a cure for cancer. We're at the beginning of a long, long road. In the mid 1950s, scientists first realized that marine organisms had great potential as a source of biomedical substances. I've been doing this work with sponges and gorgonians in Puerto Rico since 1987. We will continue to isolate and identify molecular structures until my colleagues can hopefully use some of them to fight bacteria, viruses, AIDS and cancer. There are many types of cancer, kidney, ovarian, etc. Some of the viruses today have developed a resistance to existing drugs. It is my belief that there will always be a need for new drugs in man's war against disease. The more molecules we discover, the better chance we have of finding one that will work.

BEO: You have told me of your love for the sea, that you are a diver and reef explorer. How do you collect your specimens so that you don't destroy the very reefs you care about?

AR: We use careful techniques. We take our specimens, about a half a kilogram, from medium-sized colonies. Sponges reproduce quickly. We limit what we take from each site. If we are looking for a rare specimen and there is only one in the whole reef, we leave that population alone. We dive to depths of 30-40 feet to collect the sponges and gorgonians. Again, we take only what is strictly required.

BEO : Do you like your work?

AR: I have always loved to dive and I am fascinated by chemistry. I am in the perfect profession for combining the two.

BIOACTIVE MOLECULES AND CANCER CELLS

Dr. Fernando Gonzalez is director of the Biotesting Center at the University of Puerto Rico in San Juan. He grew up on the north coast of the island, about 40 miles from San Juan. He received his Ph.D. in biochemistry from Cornell, where he also did postdoctorate work. He attended Howard Hughes Medical Institute in Worcester, MA and worked in a lab connected with the University of Massachusetts. Several years ago he returned to Puerto Rico to teach and continue his research.

BEO: Dr. Gonzalez, your colleague Dr. Rodríguez told me that you are taking his molecular compounds and using them in your work.

FG: Yes, we are in the early stages of developing a cure for cancer using compounds Dr. Rodríguez and others have identified from marine organisms. We test these compounds for their ability to inhibit growth of cancer cells using tissue cultures.

BEO: In very simple terms, can you explain what cancer is?

FG: This is an oversimplistic view of cancer: cancer cells are those which are out of control. Normally, only when there is a need will a cell divide and produce other cells to form tissue. Usually when the body is harmed a signal goes out to the cells to grow and divide. With cancer, even without a signal, cells continue to increase and divide unchecked. This uncontrolled growth places too many demands upon the body. Cancerous cells may also lose proper functionality. A normal physique is unable to keep up with uninhibited division. We are trying to find molecular compounds that inhibit this unlimited growth.

BEO: Could you explain exactly what you are doing in your labs?

FG: We are at the beginning of a long process. Dr. Rodríguez sends us compounds. We take molecular compounds (natural products) that he has isolated and determined whether they are "active" against cancer cells, that is, they have the ability to inhibit growth. We culture these cells in our lab. In further work, we investigate the molecular mechanism at work and try to see what it is in the cancerous cells that is being affected by the compounds and is responsible for the inhibition of growth. We test several extracts and in a complicated procedure try to isolate the exact compound. We call this compound bioactive.

BEO: What do you see as the end result of your work?

FG: Eventually we will be able to identify the part of the molecule that makes the compound active and find a common theme, what property generally in these molecules works to inhibit growth. Then we hope to be able to go back to the lab and modify inactive molecules so that they become more active and retard growth. Currently we have found a very promising bioactive molecule that seems to be effective against leukemia. It is active to levels that are comparable with drugs now in use.

BEO: It sounds like you are involved in a long and complicated process.

FG: Yes, for sure. To develop a new drug, once you have the beginning active formula, may take several years, perhaps decades. After the work in my lab is completed, then the compound goes to other labs and is tested on rats and larger animal systems, such as those of primates. If the compound is effective to stop growth of tumors in these animals, without dangerous side effects, eventually the drug may be used on humans.

BEO: All your work begins with corals, correct?

FG: Yes, all that we have in our current research comes from the coral world.

NEW MEDICINES FROM THE REEF

Interestingly enough, each of these four scientists is working with marine and reef creatures in the very areas close to the *Berman* spill. In fact, Dr. Santiago's research was seriously impacted when the sea urchins in the Condado Lagoon were killed off due to the oil spill. Subsequently this population is recovering, and Dr. Santiago is helping it to do so by planting fertilized sea urchins into the lagoon. Considering the significance of his life's work, the harmful ripple effect of an oil spill becomes even more apparent.

These scientists describe the complicated and painstaking processes employed by the biomedical world to cure diseases which are responsible for the deaths of millions of humans. Vital to their work are unique substances found in the sea, in organisms living among and forming the coral reefs and its associated ecosystems. The ability of seemingly defenseless marine organisms to produce toxins for self-protection has drawn the attention of medical researchers. Since the early 1970s scientists have experimented with toxins produced by coral creatures in the hopes of developing cures to man's many diseases. Not all such research is in its infancy. Many new "miracle" drugs have reached the market or are in various stages of pre-market testing. Considering that the U.S. Food and Drug Administration (FDA) requires extensive tests before a new drug can be marketed and that a typical investment into such testing takes ten years and more than $50 million, the results to date are impressive.[8]

Among the medicinal products now on the shelf or that soon will be:

- A marine *annelid* is responsible for the drug *Padan,* an insecticide effective against larvae, rice stem borers, plant skippers and citrus leaf miners.

- *Sponges* have produced:
  - *ARA-A*, an antiviral drug effective against herpes.
  - *ARA-C*, an anti-tumor drug that treats Hodgkin's lymphoma and leukemia.
  - *Manoalide*, a potentially anti-inflammatory drug used for every thing from bee stings to arthritis.
- *Gorgonians* (sea fans) produce *prostaglandins*, substances that regulate gastric secretion, help with muscle contractions and relaxation, and mediate blood platelet aggregation. For example, *Punaglandin-1, Punaglandin-2 and Punaglandin-3* were isolated from a horny coral and seem to be active against leukemia cell proliferation.
- *Sea Hares* have given us anti-leukemia compounds and are contributing to an understanding of how carcinogens function.[9]

Currently being researched are several cancer-inhibiting drugs:

- *Didemnin B* is extracted from a tunicate, a marine creature which grows as small grayish patches on coral and rocks.
- A *Peptide*, taken from sea hares.
- *Bryostatins 1, 2, 3,* and *4* from a marine animal called a bryozoan.
- *Stypoldione*, a substance found in brown alga.
- *Palytoxin* from a soft coral.
- *Punaglandin-3* from a horny coral.
- A large protein from clams.
- Shark cartilage yielding *squalamine*, a potential antibiotic that may attack drug-resistant bacteria and fights tumors.

These cancer-inhibiting compounds seem to be effective in stopping tumors in mice, increasing life span, or significantly inhibiting leukemia or other cancer-growing cells.[10]

More recently pharmacological companies and the U.S. government under NOAA's Sea Grant Program earmarked significant funds to this growing field. Other cancer curing drugs being tested are: ET 743 which seems to fight leukemia, melanoma, ovarian sarcoma and breast and lung cancers; Novascreen, which fights against nervous system disorders such as Parkinson's disease; and Pseudopterosin, an anti-inflammatory agent effective against arthritis, asthma, and psoriasis.[11]

We need new drugs from the sea because we do not have enough laboratory-produced chemicals to fight disease. Dr. Rodríguez's comment

Coral reef. (Reprinted by permission of World Wildlife Fund.)

about diseases becoming resistant to known drugs appears to be causing the scientific and medical community great concern. For years we have assumed that pneumonia, tuberculosis and other once-deadly diseases will never bother man again as long as we receive the proper vaccinations or antibiotics. But new strains of *superbugs,* bacteria resistant to antibiotic treatment, are appearing on center stage in the theater of disease. These superbugs resist penicillin formerly used for the treatment of pneumonia, and antibiotics used to cure tuberculosis, fight inner-ear infections and battle intestinal bacteria. Dangerous infectious diseases like meningitis are on the rise once again. Scientists predict the situation will worsen before it gets better, meaning more deaths and the need for alternative cures from other sources beyond our everyday household antibiotics.[12]

The wide range of drugs already on the market combat only a small portion of the many health problems afflicting man. Increasingly our proliferating populations need new cures for diseases never before known to man.[13]

We are in a desperate race with time. Man has interfered with the natural order. In the process we may be sacrificing the very beings that may hold the keys to our salvation. Many people now recognize the rainforest as "nature's pharmacy," a potentially priceless storehouse of unknown plant and animal species that may have medicinal benefit to man. The ocean's coral reefs occupy a similarly vital role in terms of biodiversity. We need a shift toward greater awareness in which coral reefs are regarded with as much value and concern for their study, preservation and protection from destruction as the earth's remaining rainforests.

Gilberto Cintron-Molero supplied the simplest argument for biodiversity with this analogy:

> *"In an airplane, thousands of rivets hold the pieces together. If one or two are missing, that might not be a big thing, depending upon where the rivet is. But what we don't know is how many rivets we can lose before the whole plane falls apart."*[14]

The earth is an immensely complex being. Mankind is but one element of the innumerable natural rivets holding it together. Coral reefs are another. How many more natural rivets can we afford to lose ... before our planet falls apart?

SECTION III

BERMAN, THE CATALYST: LESSONS FOR FUTURE PREVENTION

THE GHOST OF *BERMAN*

The *Morris J. Berman* cleanup ended officially the last day of April 1994. But *Berman* is not over. The "Ghost of *Berman*" will haunt us for years to come.

In the physical realm, oil will continue to ascend from inside the *Morris J. Berman* barge as the wreck rests on the ocean floor at a depth of 6,600 feet. As ocean waters and air warm over time, oil will also rise through the sands and from the near shore reefs to emerge at the surface of what previously appeared to be clean beaches and shoreline. Recovery of seagrass beds and other sensitive ecological areas identified in the pre-assessment damage study may take years.

On a less tangible level, *Berman* is like a stone cast into a deep pond. The ripples this spill caused on local, national and international levels will continue to reverberate in increasingly larger arenas for years to come. *Berman* is not a one-time event, limited in time and place. Like the *Torrey Canyon* and *Exxon Valdez* spills, *Berman* is a *catalyst*, shifting the world's attitude toward oil spills.

BERMAN AS CATALYST

A catalyst is "a person or thing that precipitates change." *Berman* has modified the attitudes of responders, particularly the USCG and those in the industry watching the U.S. methods for dealing with oil discharges. *Exxon Valdez* was the *wake-up call* for the U.S., bringing us out of our complacency and into some state of preparedness for oil spills in U.S. waters. *Exxon Valdez* resulted in OPA 90 and its mandates. *Berman* provides the lens through which the response community can see how well OPA 90 works, and what, if anything, must be changed to prevent and respond to oil spills in the year 2000 and beyond.

Berman serves as a focal point because the spill happened in a highly visible locale, in the heart of the tourist district of a much-frequented Caribbean island, in an important commonwealth of the U.S., and not in the middle of the ocean or in some industrialized harbor. This visibility and the staggering cleanup costs made *Berman* a test of how well OPA 90 really prepared the U.S. to respond to a spill of national significance. The gaps in response preparedness, methodology and management of a cleanup became disturbingly apparent on international, national and local levels.

Problems in the local arena were substantial. The commonwealth of

Puerto Rico has no specific authorization for spill funds and has an insufficient amount of emergency funding to activate and pay for immediate response by local agencies. There is a significant lack of spill-combating equipment in Puerto Rico. While the island has a large labor pool available to respond to a spill, few if any of these persons are trained in safe management of hazardous materials and onshore and offshore oil removal. Care facilities are insufficient to handle wildlife harmed by oil contamination.

Although Puerto Rico is a high-volume port for oil traffic from foreign countries passing through Caribbean waters, the local government has little control over inspection and regulation of the many vessels in its ports and near its coastlines. Puerto Rico must depend on USCG inspections. The significant gaps in vessels inspected and the scope and frequency of inspections performed by the USCG may well lead to a repeat of *Berman*. Although now corrected, the Area Contingency Plan (ACP) at the time failed to identify National Historic and World Heritage resources located on the island.

These deficiencies found in an economy bolstered by U.S. dollars and staffed by the U.S. Navy and USCG forces are disturbingly prevalent in other islands in the Caribbean chain. *Berman* underscores the need for efforts from the International Maritime Organization (IMO), NGOs and island governments to develop a cooperative approach to oil spill preparedness and response. This includes planning on a regional basis, establishing a coordination center, and pooling assets to respond to the oil spills that continually occur given that the Caribbean lies in the path of one of the world's oil choke points. In 1990, the world's maritime organization adopted the Convention on Oil Pollution Preparedness Response and Co-Operation (OPRC) to address this lack of regional preparedness around the globe. Implementation of this Convention, while crucial, is in its infancy.

THE SHAKE DOWN

On a national level, *Berman* provided a *shake down* to break in the new procedures required by OPA 90. The lessons learned from a response requiring 1.5 million man hours and almost $87 million in funding were staggering. OPA 90 called for the Responsible Party (RP) to manage the cleanup. When the RP does not show, this job falls squarely on the shoulders of the USCG. It became clear that the USCG did not have an on-going, trained management team equipped and prepared to take over a spill response and at the same time manage its considerable daily obligations.

Further complicating this picture is the USCG's policy of routinely changing personnel under an inflexible Temporary Addition Duty (TAD) system. The lack of trained management teams and the TAD policy are deficiencies that need a "national fix." The ACP failures pointed up a more generalized lack in national preparedness. The U.S. must reach consensus on a national level about:

- How spills will be handled.
- What systematic approach will be used.
- What areas will be given priority.
- Which roles each involved federal and state agency will play.
- How the RP will coordinate with the Federal On-Scene Coordinator.
- What management functions each will assume.

Once the national solution evolves, the response community can turn to the practical problem of incorporating consensus opinion into a Unified Command Structure with a joint decision-making group at the top, composed of the USCG, the RP and the state official in the involved state or commonwealth. The evolving firefighting incident command system (ICS) used in *Berman* has become the national model for spill response, with certain future modifications.

The USCG instituted an Incident Specific Preparedness Review (ISPR) of *Berman*. The recommendations resulting from that review address perceived gaps in USCG response. Capt. Ross (Commander at the time of the spill), CDR Stanton and other officers in the field during the incident support many of the conclusions reached by the ISPR panel, and offered refinements or additional suggestions. We must act on their recommendations and incorporate them into our national response system now and not be lulled into inaction until the next spill occurs.

OIL PREVENTION SAFETY NETS

This spill and the manner in which it happened point to a fundamental problem. Historically, the USCG and international communities believed that technological and engineering marvels could solve the problem of safety at sea and minimize marine casualties. This viewpoint expressed itself in international treaties such as SOLAS, MARPOL and STCW that focus on equipment standards, safety regulations, lifesaving, fire-fighting capabilities and design requirements. With advances in technology, these conventions emphasized engineering innovations and state-of-the-art electronics such as radar and sophisticated computer-tracking systems as the ultimate means of prevention.

Berman revealed all too clearly that the best technology cannot prevent human error. Studies show that 65 to 80 percent of all casualties are caused by people. Emphasis needs to shift to *prevention through people*, not technological fixes. Practical applications like ensuring better management, company-wide commitment to safety filtering down through all layers of a business, and well-trained and better-functioning crews focus on what people can and must do to prevent casualties.

Under the *old* system, the maritime world relied on a series of *safety nets* to maintain controls to prevent oil spills:

- The operating company – supposedly sets rules, regulations and trains its operators.
- The owner of the vessel – whose vested self-interest presumably fosters a high degree of care to avoid costly losses in operation.
- Class Society – classification societies like England's great ship insurer, Lloyd's of London, maintain a registry of all vessels at sea and sets rules for insurance purposes governing their safe operation. Ships are classified according to how well they meet the standards and their record of operations.
- Flag State – the nation which issues the operator's license.
- Port State – the nation which exercises control over the safety condition of vessels in its waters. For example, the USCG has the right to board any foreign vessel entering U.S. waters and inspect that ship for compliance with safety provisions and international treaties.

In theory, these safety nets reinforce each other and protect the ultimate consumers and the environment as the final recipient of released cargo. If one safety net breaks, the next net in line should prevent a discharge.

HOLES IN THE SAFETY NETS

In reality, the safety nets of the past are failing. A reorientation of the entire system is in order. A study of *Berman* reveals accelerated changes in the shipping industry and the shifting perspectives of those charged with ensuring safe operations. Instead of arraying regulatory agencies against companies and their insurers, the intent of the new process is to bring vessel owners and operators into willing cooperation with the safety requirements.

Model systems are evolving on national and international levels that stress reward for *good guy* companies and shrink the sphere of *Mom and Pop* (one-ship companies) or shady operators. The goal is to eliminate substandard shipping in U.S. and foreign waters by participation at all levels. IMO and the International Petroleum Industry Environmental Conservation Association (IPIECA) are sponsoring a series of seminars designed to focus on preparedness and prevention along the major global oil routes.

IMO adopted a resolution implementing a code for ensuring better safety of operations (ISM Code). This Code will now be incorporated into SOLAS and become effective July 1, 1998. Many companies such as Texaco have instituted internal auditing procedures for their national and international subsidiaries to detect problems and correct them before a spill occurs.

The U.S. is taking a lead role in spill prevention. While the phase-in of *double-hulled* vessels by the year 2015 is a necessary technical move to avoid future spills, the USCG, the primary maritime regulatory agency, is shifting its focus. *Berman* accelerated this radical shift from reliance on technical standards to *prevention through people*.

In response to the growing receptivity at national and international levels, the USCG and subcommittees of IMO instituted a series of studies and work sessions to address broader issues. These included such matters as:

- Control over unregulated vessels such as towing vessels.

- Regulations providing for stricter and more practical licensing requirements for operators of all types of vessels transporting oil and hazardous materials.

- Incorporating radar and other high technology equipment into ships with proper simulator training in their use.

- Emphasis on improved training as a way to partially fulfill the current license requirement of hands-on experience. While practical experience is a must, there has been little quality control over actual *sea miles*.

The Towing Vessel Safety Act was introduced in the 1994 session of Congress as proposed legislation that incorporated certain solutions to address the above concerns. Unfortunately, the legislation failed to pass. This has not stopped the USCG from attempts to achieve their objective of control over this unregulated industry. The USCG is proceeding by way of the regulatory process to improve current standards.

The U.S. has developed a *point system* for inspecting foreign vessels entering our waters. This system affects a company's bottom line. Those vessels rated poorly will find themselves boarded time and again on entering U.S. ports. Compliance with safe operational standards should become less costly than the previous mode of many operators who cut costs and defer repairs. Now operators may suffer adverse economic consequences for cutting costs. Their all-important shipping schedules will be blown out of the water through frequent boardings, inspections and detention. The USCG is starting to adapt this same point system to non-complying or historically poor U.S. operators like the Berman/Frank family.

Emphasis is increasing on international *Port State Control* so that countries can control ships visiting their waters where the *Flag State*, the country of the ship's origin, fails to do so. To address a Flag State's omission, IMO amended older treaties (specifically the STCW, expected to be in force in February 1997) to exercise greater vigilance over Flag States who issue certificates of operation to vessel operators. The treaty amendments should ensure that vessels comply in fact with the safety standards. When absolutely necessary, countries can resort to the Intervention at Sea treaty which permits them to board a ship and take control of the vessel and its operations if the ship becomes or presents an imminent threat to a country's waters.

GLOBAL FUROR

Berman is a focal point because of its timing. This was the first U.S. spill of significance since the effective date of OPA 90, August 18, 1993, and the first one in which the USCG assumed the role of RP. *Berman* provided the chance to see how OPA 90 plays out in a real life scenario. In addition, at the time of *Berman*, the U.S. had enacted certain legislation that reflects our increasingly aggressive posture toward polluters of national waters. This new legislation shifts responsibility to the party causing the spill (OPA 90) to ensure that there is sufficient insurance to cover the costs of cleanup, and to fully restore or replace all natural resources affected, including placing a reasonable valuation on their interim loss use or paying for other mitigation activities.

The steep cost of *Berman* highlights the U.S. *polluter pays* doctrine. OPA 90 differs from the international conventions and funds by exposing the shipowners, RPs and their insurers to unlimited liability for oil spills. The new Certificate of Financial Responsibility (COFR) is effective for tanker and barge owners now and for all vessels trading in U.S. waters by December 28, 1997. The COFR has the potential for making the insurance industry unlimited guarantors for their clients, the operators of these vessels.

The range of damages for natural resources is being extended by the latest NOAA regulation concerning Natural Resource Damage Assessment (NRDA) for damages caused by a spill. While the international community limits recovery to economic damage only, the NOAA regulation provides for use of *contingent valuation* (CV) and computer models as a method for assigning a value to the non-use losses of natural resources. The international community regards CV as a particularly onerous concept because of the use of non-traditional means for valuing resources without a clear market value. The U.S. regulation requires the spiller to restore the injured resources to the public or to replace them with equivalent services both human and ecological.

These regulatory efforts created a global uproar that left the U.S. in a position of being the staunchest world defender of clean waters and NRDA. America is leading the international community well beyond what it would otherwise be willing to do to address the marine oil spill problem. The ultimate impact on the shipping community and marine casualties from these regulatory efforts is as yet unknown, but it appears to be environmentally beneficial. Regardless of worldwide opposition and response, the U.S. seemed intent on standing behind its vision of oil-free oceans.

Yet the overall U.S. position is subject to shifting political currents. Recent changes in the makeup of Congress may change our position as a world leader. *Berman*, unfortunately, may be a watershed. The latest attacks in the U.S. Congress seem determined to swing the pendulum back to limiting how much the polluter pays. Public reaction to the severe conservative backlash already indicates significant voter disenchantment and underscores the need for continued vigilance and awareness as the tide on these issues fluctuates.

THE FUTURE HOLDS PROMISE

Scientists, NGOs and governments on the national and international level are uniting to study, monitor and better protect the world's dwindling natural resources so dramatically affected by our insistence on oil as a primary energy source. Their tools are marine protected areas, biosphere reserves, specially designated areas, world heritage sites and a host of other designations under international treaty and national and local law. The U.S. and seven other governments instigated the International Coral Reef Initiative in 1994 which the international community, World Wildlife Fund (World Wide Fund for Nature) and many other NGOs throughout the world now support. ACEC is the key:

- *Assessment* through monitoring projects.
- *Community Involvement* in protective projects.
- *Education* about the wonders of the marine world.
- *Commitment* to long-term programs designed to protect fragile coastal ecosystems.

The International Year of the Reef 1997 was launched at the 8th International Coral Reef Symposium in Panama City, Panama in June 1996.

ON THE TIP OF INTENTION

Berman is a metaphor, a glaring symbol of an instant replay of oil spills occurring daily throughout the world, and of the U.S.'s determination to stop the flood of oil into the seas with effective and hard-hitting legislation. We live in a relatively small global society. Decisions made in New York affect an island in the Caribbean. Policy shifts from USCG headquarters in Washington, D.C. impact the Philippines, whose seafarers constitute one of the largest work forces for oil shipping countries throughout the world.

Oil discharged from a spill in Puerto Rico eventually reaches the earth's polar ice caps. People who depend on and demand oil for energy caused OPA 90 to exist. Along with the privilege of enjoying this supply of energy fuel comes the responsibility to guard and protect our oceans against degradations. As responsible citizens, we must keep the U.S. focused and committed on the right track. Our insistence that we remain leaders of global pollution policy may avoid serious backsliding nationally and internationally.

One government alone, no matter how powerful, cannot stop the daily poisoning of the earth's waters. After all is said and done, only individuals make the difference. A basic philosophical concept holds that our intentions affect our actions, which in turn change the events of the world. *Everything rests on the tip of intention. Berman* may be the catalyst, but necessary change will occur only if there is a collective shift of attitude. In order to protect the living kingdom which is essential to our own well-being, it is vital that we *all* become "good guys" on the petrohighway.

For this to occur, the motivation behind our actions should stem from a real desire to *clean up our acts* and prevent having to react with massive cleanups like *Berman*. The last chapter will offer more specific suggestions about things each one of us can do to improve the current situation. Sometimes a relatively simple act such as shifting to phosphate-free laundry detergent can have enormous ramifications for what flows into the sea and affects coral reefs. The national awareness and participation in recycling programs over the past decade is an encouraging example of what can be accomplished.

On a more politically active level, contact your senators and congressman and express your support for important legislation that strengthens towing vessel regulation, mandates adequate water quality standards, protects wetlands and maintains the *polluter pays* concept of OPA 90. Your voice resonates as a constituent on the telephone or in a letter or fax, and you will gain the eye or ear of your elected representatives. Talk to your friends and encourage them to do the same. Imagine the potential impact of every U.S. citizen taking this one small step. This is how individuals change the world.

The effect of these acts may help prevent the prospect of artificial concrete reefs of the future, pale imitations which can *never* come close to duplicating, substituting for or replacing the abundant life of a healthy, flourishing coral reef. Then *Berman* oil balls can cease fouling beaches and rocky coastline. Clean ocean waves can again grace the world's seashores. The result is practical as well as aesthetic: the next $87 million we save by not having to cleanup another *Berman* can be used in so many more productive ways to benefit mankind and the creatures who inhabit this planet we share.

RESPONSE MANAGEMENT AND LESSONS LEARNED

"AN OUNCE OF PREVENTION"

In many ways, the *Berman* response was a success. Of the 1.5 million gallons of oil on board the *Morris J. Berman*, responders accounted for some 1.3 million gallons. Of the 1.5 million hours worked by thousands of individuals, only 15 reportable OSHA safety incidents occurred, with all but one involving only minor injuries.[1]

Even so, deficiencies exist at all levels of the response structure and threaten the success of this and future spill cleanups in U.S. waters. The oldest adage in the response business is, *"An ounce of prevention is worth a pound of cure."* Prevention is the first line of defense. But once the oil is out of the tanker and into the sea, the next line of defense is quick and effective action. The U.S. and international communities still have a long way to go in both the prevention and response ends of the oil spill business.

The problem is perhaps best exemplified in the ongoing debate at national and international levels about the most effective method to prepare for and then respond to a discharge. The U.S. is groping toward a national system of response. Unfortunately, we no longer have the luxury of resolving our dilemma in isolation. We are part of the global society. International decisions affect U.S. waters. Decisions or regulations in U.S. waters affect shipping worldwide. Unless and until there are answers worldwide to these questions, overall effectiveness of spill response remains in jeopardy.

We live in an age of increasing environmental awareness. The public now demands that governments and responsible parties act quickly and efficiently once oil is spilled. Consumers are no longer willing to sit back and watch the "spill of the month" happen with impunity. This reaction found its voice in the U.S. with OPA 90 and its many requirements. OPA 90's primary response requirement states that *"Owners and operators of vessels must have pre-approved response plans to deal with a possible oil discharge in U.S. waters."* These plans describe the organizational structure and the type of spill management team that will spring into action when and if a disaster occurs. Concurrent with the vessel response requirement is the National Contingency Plan system which has created national, regional and area response committees. The job of these committees is to formulate plans for each state, commonwealth and territory of the U.S. to deal with marine discharges.

On the international level, the International Maritime Organization (IMO) has successfully produced two conventions now ratified by sufficient countries to make them effective. In a move similar to what the U.S. has achieved through OPA 90, IMO sponsored Regulation 26 to MARPOL, one of the strongest marine treaties in existence, requiring vessel owners to have in place a shipboard emergency plan for oil spills. Regulation 26 applies to all oil tankers of 150 gross tons or more, and to every other ship of 400 gross tons. All vessels must have such plans in place.

The second step taken by IMO to further world response was passing the International Convention on Oil Pollution Preparedness, Response and Cooperation of 1990 (OPRC), effective as of May 1995.

INTERNATIONAL SPILL PREVENTION AND READINESS: THE OPRC STORY

Recognizing that prompt action is the only way to properly deal with a major discharge, the OPRC treaty establishes a framework for international cooperation and mutual assistance. In the event of a spill, either on board a ship or at an onshore facility, the contracting parties agree to report that incident to their nearest neighbors and to IMO, giving countries closest to a spill advance notice to prepare as future recipients of the discharge. Where a country has insufficient resources, its neighbors can provide equipment and personnel to be reimbursed later by the requesting party.

The second major thrust of OPRC is to impose an obligation upon its signatories to create a national system for responding to spills, requiring at a minimum that each country has in place a national contingency plan for handling the response, pre-positioned oil spill equipment, response organizations and plans of communication and coordination between those involved. This requirement brings the international community into step with the U.S., which has long required such capabilities on a national basis.

IMO developed a number of tools to encourage implementation of OPRC requirements and to aid signatory countries. A 1991 IMO Working Group composed of experts from various NGOs and governments produced guidelines for dispersant application, response facilitation, model courses on preparedness and response, and field guides for tropical water oil spills. IMO organized a coordination center staffed by French, Japanese and U.S. personnel to assist in educating, training and providing technical support to requesting countries.

In a creative move, IMO established a cooperative venture with industry to encourage preparedness for discharges. The International

Petroleum Industry Environmental Conservation Association (IPIECA), a major oil industry organization, joined as a partner with IMO, the United Nations Environment Program (UNEP) and many others to establish a series of workshops on spill response planning beginning in 1993. These symposia were intended to help countries in each oil-sensitive region develop plans and establish mutual cooperation pacts with each other to respond to discharges. Regional meetings have been held in the Arabian Gulf involving seven member states; in the Mediterranean involving Egypt, Israel, Cyprus and all states bordering the Mediterranean and Red Seas; in the Caribbean, involving 22 island states; in Southeast Asia, involving Japan, Malaysia, Philippines, Singapore, Thailand and others; and in Latin America, including Brazil, Chile, Peru, Uruguay, Venezuela and other member states.

The success of these meetings is evident. In each region, government and industry joined together to produce regional centers, stockpiles of equipment, coordinated response plans and assistance sharing to promote quick and effective response to spills.[2] An innovative project bearing the acronym GOSPIL, short for Global Initiative to Enhance the Capacity of Countries to Prepare for and Respond to Marine Spills, may achieve dramatic results when put into operation. This initiative would put $30 million over five years into projects like spill response centers: $5.25 million would establish three regional oil spill centers, $18.5 million would be used to develop model training courses in over 50 countries, and $2 million would be earmarked for workshops.[3]

Unfortunately, progress in this area may be delayed. IMO recently voted for a much lower budget, one that barely keeps up with inflation and which eliminates permanent positions such as coordinator of the new spill centers, without whom the centers are meaningless. This budget crunching will frustrate certain OPRC effectiveness proposals, like developing the International Ship Information Database (ISID) which would contain safety records, inspection data and insurance information on vessels worldwide. The cutbacks will leave vacant certain needed personnel positions and cause reductions in training programs.[4]

NO MAGIC FORMULA

The goal of OPA 90, Regulation 26 and OPRC is to produce a *Response Management System* (RMS) that works. Unfortunately, there is no magic formula to help designers create a perfect response system. Industry consensus about what such a system should be does not yet exist.

The response community is caught in an uncomfortable tug-of-war between two competing principles: *Effectiveness* and *Efficiency*. *Effectiveness* is doing the right thing (or getting the right thing done). This includes safe response, preventing further oil leaking into the environment, recovering the most oil possible, and minimizing impact regardless of the economic value of the affected natural and other resources. Effectiveness measures results.

But effectiveness often conflicts with efficiency. *Efficiency* is doing the right thing correctly or getting the right things done *with the right amount of resources*. Often efficiency translates into using the most conservative or minimal amount of people and equipment and no more, getting the most out of the resources on hand before importing other means, keeping the scale of response in proportion to the size and relative importance of the affected environment, and balancing cost with benefit.

According to researchers for Scientific and Environmental Associates, Inc., the clash happens when governmental entities, environmental groups and public officials push for an effective response, while company management, Prevention and Indemnity (P & I) clubs and insurers argue for efficiency.[5]

This is not a new battle. Bottom line, *reasonable* thinking and commitment to the cleanest environment possible compete across the board in all areas of society. The answer may lie somewhere between the two principles. *Ease of execution,* with its emphasis on people, may become the goal, with people, not principles, making response work.[6] Oil spills are highly emotional events. The FOSC is terrified of "botching it." Politicians are worried about re-election if the cleanup does not go well. The company at fault faces possible bankruptcy, and its board members and officers face jail time. Contractors doing the cleanup work are concerned about being paid. The marine wildlife are trying to survive the nightmare.

Amid all this chaos, the emphasis should be on applying *people power* in the safest manner designed to achieve an acceptable outcome. The conflict focuses on what constitutes an acceptable outcome. The U.S. view suggests that cleanup should be performed as cheaply as possible, but not at the expense of conducting an effective response. The costs of preventing spills and responding to those which occur are costs of doing business, and like wages, taxes, and other similar costs, must be passed on to the consumer. The fact is that spilling oil is expensive.[7] Oil spills demand flexible approaches. What works best in a specific situation dictates, rather than some fixed policy.

No matter how the debate about response and preparedness is resolved, the refrain underlining this entire issue is *prevention*. According to Dr. Jacqueline Michel, an expert in the field of oil spill response, a member of NOAA's scientific support committee, and Director of the Environmental Technology Division of Research Planning Inc.:

> *"Even under the best of conditions, oil spill response is only marginally effective, once the oil is out of the vessel. A 'good' response*

recovers 20 percent of the spilled oil. That's a lot of oil left in the environment. So, improved response preparedness is only a better band-aid. The real benefit of OPA 90 has to come from spill prevention. Keep the oil in the vessel."[8]

CLOSED vs. OPEN SYSTEMS

Complicating this unresolved conflict in competing principles is a difference in response styles. There are basically two types of operating systems for managing oil spill response: *closed system* and *open system*. Europeans and much of the rest of the world seem to favor the *closed system* approach. This is a more traditional military model, where control rests with individuals at the top with little or no input from political, economic and cultural entities involved in the actual spill. In most international communities, the appropriate governmental entity handles spills, with local jurisdictions managing limited impacts on coastlines and inland areas.

Nations choosing to manage their own cleanups reason that most spills occur from vessels passing through their waters, what the industry terms the *passing ship syndrome*. There is little or no Responsible Party presence in the victim country. The logistics of spill response enable domestic personnel rather than a RP passing through to react more quickly and to better handle the discharge.

This type of closed system occurs in the U.S. in certain situations, such as firefighting. Suppose there is a fire in your hotel. The best firefighters respond, directed by a fire chief. He does not bother discussing how he intends to handle runoff from water sprayed onto the burning building with the local environmental group. His job is to put out the fire and save lives. The fire chief knows the laws under which he works. While the fire rages, he is in charge.

In contrast, the *open system* is a problem-solving approach that depends on internal feedback from members within the same group and externally from all concerned stakeholders. *Berman* involved 15 federal and state agencies and 50 contractors. The positive aspects of an open system are that decision-making is aided by input from all concerned, greater flexibility is possible during an event, and everyone benefits from feedback as different solutions are attempted.

The problem with open systems is that too many cooks spoil the broth. This more flexible approach works only when all concerned stakeholders are identified in advance of the crisis, agree on common goals and a shared model of how a response should work, and receive appropriate training to oversee their various assigned functions. Integration of every part into the whole is crucial. If one part is missing, the entire response may collapse.

It remains unclear which type of system the U.S. will use to respond to oil spills. The U.S. is in the process of adapting a system that best suits a democracy in which the military, in the form of the USCG, may run the show. An Incident Command System (ICS) is emerging that is more open than closed. The 1994 revisions to the National Contingency Plan call for some form of unified command system. The RP and FOSC share the top spots. State and local agencies involved have substantial input in the decision-making process. The area committees adapt this mixed model to the peculiarities of their local environment. In other words, there is a standard methodology with the area committees working out the details peculiar to their own needs.[9]

"HEADLESS CHICKEN"

The U.S. chose to involve the RP and others in spill response, a different path from the international community. Our country reasons that OPA 90 gives companies strong incentives to better handle their own discharges. Under OPA 90, failure to respond subjects the RP to unlimited liability for all consequences of the spill, as well as possible criminal sanctions for individuals in the company. Given these factors, the U.S. contends that the RP can and will better manage all aspects of the spill response from start to finish. To ensure careful response, the EPA or USCG stand by and monitor the RP's activities. If the RP fails to show or assume responsibility for the cleanup, one of these agencies assumes a more active role beyond merely setting procedures and standards and overseeing outcomes.

Critics roundly declare that the U.S. system does not work. According to Joe Nichols, ITOPF:

> "Quite simply, spill response systems outside the U.S. are better and more cost-effective. The RP is not responsible for organizing the cleanup. Instead, we rely on one or two lead agencies to manage. This avoids the confusion and disaster that so often characterizes U.S. responses. Most spills I've witnessed in U.S. waters remind me of 'headless chickens.' There are too many people involved."[10]

The French level similar criticism at the U.S. approach:

> "In many parts of the world, there are a series of events which make closed systems work. For example, there is the problem of the 'passing ship syndrome' where ships simply move through the waters of Western Europe, Africa, or the Caribbean and don't originate from or stop at a country's port. If an incident occurs in the waters of one of those countries, you can't expect the polluter to take responsibility for fighting the spill. There are various reasons in the

U.S. why the open system works, including groups that are used to and know how to work together. But the U.S. is not the rest of the world. The OPRC provides the framework for a national response system. It is designed to serve the nations in a particular region. Don't export your system."[11]

Members of IMO also question the U.S. system:

"The U.S. is in the forefront in awareness of pollution, equipment, and combating spills on a technical basis. But America is way out of step with the international community in terms of response and compensation. Not all countries have the means, equipment or money to fight spills. That's why the OPRC is so important. It emphasizes regional cooperation, allowing equipment and people to pass across country borders during an emergency and dealing with the accounting issue later. We'll be using a regional center, for example in the Caribbean, as the base of operations, coordinating the cleanup from that center. What is needed is a more universally accepted scheme that the U.S. is a part of."[12]

There are those within the U.S. system who object to reliance on this hybrid open system:

"We in the U.S. are sending mixed signals with our form of Response Management System. According to OPA 90, we believe that 'the polluter pays' and we hold the polluter responsible for the response. Yet the FOSC is the one in charge. This leaves the RP in a position of always worrying about being whacked on the side of the head. While the RP is supposed to be in the lead, he isn't. What we need is to be clearer about who's making the decisions. The USCG is there to help, but the polluter in partnership with us needs to make responsible decisions."[13]

A consensus emerges: the U.S. needs to decide on a national approach to spill management, pick either an open, closed or mixed system, and then fashion response action accordingly.

SIX CRITICAL FACTORS

Regardless of which system is used or what guiding principle controls, experts agree on six critical factors that make for a successful RMS:

1. *"The salvage operations must minimize spillage of oil ... and must not interfere with pollution response operations."* In other words, the first

goal is to get rid of the source of pollution without interfering with the cleanup.

2. *"The immediate response by the RP and the Coast Guard must mobilize enough appropriate response resources (people and equipment) to contain most of the oil at or near the source and to protect sensitive areas."* Protect the environmentally sensitive areas wherever possible.

3. *"The response organization must be able to communicate and manage information internally and externally."* Those involved need effective feedback so they can modify their strategies and tactics, adapting to the situation and applying what works.

4. *"Coordination between federal, state and local organizations and the RP must be pre-planned, account for stakeholder interests and ensure a response organization that will be cohesive and effective."* Pre-event planning is crucial. Bring everyone together in one room ahead of time.

5. *"The response organization must be capable of sustained effective operations."* Spill response is a many-headed hydra. Responders must act in coordinated fashion in many areas over time. Rotation of people over time can be a major impediment to effective response.

6. *"The response organization must meet the public's realistic and achievable expectation for pollution response."* Zero environmental impact is doomed to failure. Reasonable expectations, not a restored pristine environment, are what responders should aim for and achieve.[14]

ISPR: REALITY CHECK?

Berman helped move the U.S. along the path toward defining the country's response system. The USCG commandant in Washington, D.C. initiated for the first time a critique of how the USCG responds to a spill. He convened an Incident Specific Preparedness Review (ISPR) to conduct a *reality check* to see if gaps existed between planned and actual response. Given the novelty of this endeavor for the higher command, and the *hurry up* methodology used by the team assigned this task, some within the system believe the report to be fraught with factual misstatements. In spite of these supposed inaccuracies, ISPR identified some key problems inherent in the U.S. response system, and proposed solutions:

- The Area Contingency Plan (ACP) failed to tailor the model to specific needs. Solution: develop a skeletal USCG organization, with area committees fleshing out the detail. Use a quasi-military open system.
- The USCG lacks a specialized response management team. Solution: The USCG should form its own spill management team,

composed of units for handling logistics, finance, monitoring and operational areas.

- There was little *interoperability* between the Vessel Response Plan and the ACP. Solution: coordination between plans. Salvage must be addressed in any Vessel Response Plan.

- Local USCG forces are insufficient to handle a significant spill. Solution: such forces should be standardized with set teams for managing each aspect of the spill. Personnel cells must be put into place quickly at the spill locale to feed and care for the additional people.

- The FOSC did not utilize the full capability of special forces like the National Strike Force, NOAA and SUPSALV. Solution: better communication about each force's capabilities and resources. The people in these organizations are full-time oil spill specialists who can bring expertise to the response.

- The USCG lacks a good system to manage information gained during a spill. Solution: create a system to retain lessons learned.

- Resource accounting needs to start on Day One. The maxim "Shoot first, ask questions later" should be modified slightly to "Shoot first, count your bullets, and ask questions later." Not all jobbers should be employed on a *time and materials* basis.[15]

This is the official report using 20/20 hindsight and after-the-fact review of a very tumultuous and rapid response to a potential major disaster over a three-month period. Others suggest a different approach to the future of U.S. oil spill response. Two individuals most directly involved in *Berman,* Capt. Ross and CDR Stanton, state their overall concept, given the realities of applying a quasi-military approach to a cleanup performed primarily by nonmilitary contractors. Said Stanton:

> *"The USCG is not staffed at the Marine Safety Office level to manage a major spill. That's the job of the Responsible Party usually. The problem is where the RP doesn't show up, as happened in* Berman, *and you have a significant spill. Where the USCG handles the response "soup to nuts," we need to do things differently. If the FOSC is to assume full responsibility, he has to have a management team ready to go, trained and able to help him. The temporary USCG personnel, TAD's, can't rotate out like they do now, every two weeks to 30 days. Their coming and going just creates turbulence. They have to be assigned to the scene for the duration. Or the National Strike Force Coordination Center can provide the management team*

for handling the big ones. Or the USCG can pre-contract and hire out a management team. I don't see any way around this problem except to exercise one of these three options."[16]

BE LUCKY, BE LUCKY, AND BE READY

Capt. Ross lived and breathed the *Berman* spill every day and night of the event from January 7, 1994 to its conclusion. Ross graduated from the USCG Academy in 1973 with a Bachelor of Sciences degree in ocean engineering, then earned an M.S. in systems management from Florida Institute of Technology. He was the commanding officer of MSO San Juan during the *Berman* spill. Capt. Ross is a highly qualified individual whose opinion is valuable. Other MSO officers may have been qualified to handle the *Berman* nightmare, but Capt. Ross proved himself the right man at the right place when the time came.

Capt. Ross shares the concerns of the ISPR panel. He is aware of and listens carefully to criticism from Joe Nichols and the international oil community. He has endlessly reflected on the *Morris J. Berman* spill response and subjected himself to closer scrutiny than any outsider could ever place on him. Capt. Ross has reached some interesting conclusions about how the U.S., and specifically the USCG as managing on-scene coordinator, should develop plans and respond to oil discharges. While Capt. Ross's thoughtful critique in his report "The Response to the T/B *Morris J. Berman* Oil Spill" (August 1995) covers some seven major areas and 26 minor points, he focuses on the key problems:

> BARBARA E. ORNITZ: What is the key to a successful response?
>
> *CAPT. ROBERT ROSS: There are three rules for any oil spill response: "Be lucky, be lucky, and be ready to take advantage of your luck." In* Berman, *for example, we had immediate notification. Two men were on board a supposedly unmanned barge. They weren't supposed to be on the barge. But because they were there, we received a call minutes after the grounding, a little before 4 a.m. My office is very close to Punta Escambrón. We had boom in a van ready to go almost immediately near the sight of the grounding. Crowley had trained personnel we knew and could contact right away from our 'drill' with the* BGI Trader *spill three weeks before. All this luck, combined with our state of readiness, meant we were able to put our boom into the water about 10 feet ahead of the leading edge of the oil heading into the San Antonio Channel and the cruise ship piers.*
>
> BEO: You're not suggesting leaving everything up to luck.
>
> *RR: Of course not. You need good pre-event planning and a flexible response system.*

BEO: What is your view about the debate concerning the type of Response Management System the U.S. should use, a closed or open system, an ICS approach, or what?

RR: We in the U.S. are not yet ready as a response community. We are far better prepared in terms of equipment and many aspects of logistics and operations, but we don't have a coherent approach towards managing oil spill events of this magnitude. We aren't 'singing off the same sheet of music' yet. What we need to do is to adopt the organizational and management principles which underlie the military general staff model. You identify all the tasks that need to be done to ensure success, and then you divide the work up into coherent bite-size chunks. This is basically where the ICS model comes from. The original military staff model was adapted to fit a different operational problem, but the basics are the same.

BEO: How does this translate to Response Management?

RR: The U.S. needs to agree on standardized operating procedures in the field, develop a tactical response manual for managers at the scene and behind the scenes, use the area committees to do our local planning, and train all those involved about how to work together to respond effectively. When a Spill of National Significance (SONS) happens, we'll know who the players are, what each one's role will be, and we can respond with preparedness and confidence. You don't want to be storming, working through the details, in the midst of a response. You want your contractors to know how to work together, know who's in charge and what to do, and not be learning "on-the-job" during the emergency.

BEO: How do you propose conducting Response Management? For example, do you favor the ICS model?

RR: I don't bow down to the ICS god, if that's what you're asking. The ICS model was developed for situations where there is no villain, where you essentially 'send the bill to God.' In the usual ICS application, no one faces criminal prosecution or financial ruin. ICS is the right model to start from, but it requires tailoring to fit the oil spill scenario. Too rigid an application of ICS won't work for oil spills. The RP and many other organizations need to be involved with the performance/involvement expected from them clearly spelled out. What we should do is draw on the strengths of this system while tailoring it to the oil spill scenario. We identify in advance what needs to be done, divide up the responsibilities into appropriate categories, set up the organization where everyone knows what their role is, and how the system functions. Then we adapt this model to the individual characteristics unique to each spill.

BEO: Can you give me a practical example?

RR: On the most basic level, when it was obvious the USCG would have to manage the spill, I had to contract with the various response companies so I could keep the spill response going after the RP's insurance ran out. I did not have access to an adequately sized and equipped Coast Guard emergency contracting, financial management and cost documentation capability that I could activate with a phone call. I did the best I could, made the agreements I needed, and decided to worry about the paperwork later. Response effectiveness, not administrative ease, was my main concern. Now I'm facing a lot of problems because of lack of paperwork that could have been avoided easily if the Coast Guard had a national capability in place.

What I would like to see is a service level capability, housed at the National Strike Force Coordination Center or perhaps elsewhere, which is "spring-loaded" and is ready to provide the necessary contracting, financial management and cost documentation support wherever and whenever it is needed.

BEO: How do you respond to the ISPR suggestion that the USCG should "shoot first, account for the bullets, and ask questions later?" I'm most interested in your feeling about accounting for cost, particularly in light of the international view that the USCG does not control costs to any degree of reasonableness.

RR: First, let me answer the question about accounting for our bullets. Where the RP handles the spill, the RP is in charge of this department. When the RP defaults, the USCG has to deal with this problem. I don't care who does the bookkeeping. It's a job that needs to be done. The USCG can contract the problem out. I'm only concerned that the spring is loaded before a spill happens, so that we know who handles accounting for cost and there is a "method to the madness."

ISPR visited San Juan right in the end stages of the response. The accounting information was available, not yet in finished product form, but every cost was documented. My larger concern is with the cost of Berman *and the misunderstandings that have centered around that cost. Oil spills are costly. How much they cost is not really known. Most responses are handled by RPs and you never see their numbers.* Berman *was a USCG managed spill with all information released publicly.* Berman *dealt with a highly persistent oil in large quantities, in a physical environment where the oil had to be removed in order to minimize environmental, property and economic damages.*

If you compare the Berman's *dollar per gallon cost to that in other large removal operations involving high recovery rates of heavy black oil, if you compare apples to apples, you will see that the* Berman *costs are not way out of line. In fact, they compare fairly well.*

Another issue which is most important is the myopic focus on cleanup costs as the sole measure of response. The objective of response is to minimize the total negative impact of the event, including not only the cost of cleanup, but impact on property, public safety and the environment. In the absence of an effective response, NRDA damages, for example, may later dwarf the cost of cleanup. Money spent on response may well pay huge dividends in the form of environmental damages avoided.

There is a point where you have to decide whether you spend twice the amount of money to get 97 percent effectiveness, or you spend half that amount to get 95 percent. We evaluated many of our decisions in terms of common sense when conducting the cleanup. For example, I thank God the Puerto Rican authorities allowed us to discharge the water from the swimming pool and smaller stilling pools into the ocean, where it was almost clean water and any remaining hydrocarbons would break down over time. Certain of the states would not have allowed us to discharge any reclaimed water, no matter how much sense that would have made. That kind of unreasonableness would have added unbelievable cost and difficulty to the cleanup."[17]

A SUCCESS STORY?

Was the *Berman* response a success? It is if one applies these critical factors:[18]

- The source was contained as quickly as possible, given the weather and difficulties of removing oil from the leaking barge. By Day 9, the *Morris J. Berman* was floated and sunk 20 miles offshore.

- The immediate response mobilized sufficient resources to contain the maximum amount of oil near the source. The San Antonio Channel was boomed. Other sensitive resources, including cultural resources, received quick and efficient protection. Three months after the spill, one of the most heavily impacted beaches, near the Hilton Hotel, was opened and judged by even the harshest critics – the press – to be "clean."

- All major stakeholders were identified, were able to express their concerns, reached mutually acceptable solutions, and pulled together in teams. They met together daily in one room, voiced concerns, decided on a plan, and took action. There were problems of logistic coordination. Given the huge amount of equipment and personnel on scene, these hitches were insignificant.

- The response was carried out over three months. When a second major release occurred, affecting areas northwest of the initial site, teams were dispatched quickly and well coordinated. When the job of one agency or contractor was completed, the *ramp down* phase was conducted in a fairly reasonable manner.

- The public response was mostly favorable. The general public seemed satisfied that all concerned were doing everything they could. Tourism is a big industry in Puerto Rico. Even with the spill, the 1993-1994 year produced almost $1.7 billion tourist dollars, more dollars than were generated in the previous year.

Regarding the cost issue, comparison of the *Berman* response with other major spill cleanups puts *Berman* on a dollar per gallon ($104.00 per gallon) spilled basis at the middle end of the spectrum.[19] Another study disputes this statement, placing *Berman* at the higher end of response costs, with barge and tanker cleanup costs per gallon in 1993 dollars running between $1.00 per gallon to $139.36 per gallon, excluding *Exxon Valdez*, which cost $276.33 per gallon.[20] (See chart page 42)

Capt. Ross disputes the claim that *Berman* was a high-end cost response, arguing that the oil in *Berman* was persistent oil, increasing the costs of cleanup, that other U.S. spills do not require the logistics of transporting equipment and machinery thousands of miles, and then housing the work force at tourist rates, and that the enormous potential for damage to natural resource and economics in the spill must be factored in. [21] Regardless of which figures are used, *Berman* was no bargain. Still, in the opinion of almost all of those interviewed, the *Berman* response was considered a success.

THE FUTURE

When all is said and done and all recommendations are taken into account, one inevitable human trait remains that may unhinge the entire Response Management System. The response business is highly cyclical. We tend to react strongly to crisis and disaster. Groups meet, reach consensus, start working on solutions and direct funding, energy and time to preparedness for the next disaster, or to avoid the next discharge. Then, over time, the nightmare fades and preparedness no longer receives top billing due to a more immediate crisis of the month.

Fortunately, oil spills the size of *Berman* are rare events. Unfortunately, they are not rare enough, as the 828,000-gallon oil spill off the Rhode Island coast near a major ecologically sensitive area in January 1996 testifies. The response community recognizes a disturbing trend after each of these significant spills. When the crisis is over, everyone returns to business as usual. The challenge for the future will be for the U.S. to sustain momentum in preparedness absent the catastrophe of the moment.

Partnerships are the key to an effective Response Management System. If the partners, those companies, governments and agencies involved in response, revert to wrong bottom-line thinking once the nightmare of a spill cleanup passes, the best laid plans will be frustrated. Cuts in prevention and training, from a short-term perspective, may appear right for business – until the next spill. This mentality is historic. It is what led to the *Exxon Valdez* response problems, when most of the spill response equipment was below deck, undergoing casual repair.[22]

As USCG Rear Admiral James Card says, *"Oil has to go first class, not economy class, and not business."*[23] This equates to responsible transportation and storage, partnership of all those working toward the mutual goal of preserving a quality marine environment, and continuous, not cyclical preparation.

The time, people power and cost expended in the *Berman* cleanup reaffirms the adage that the first line of defense is prevention. We should insist that nationally and internationally the focus on response and preparedness continues with the same intensity, well after the memory of *Berman* fades. If so, regardless of the response systems applied, we will benefit. If not, ultimately the consumer will pay the price, and the ghosts of *Torrey Canyon, Amoco Cadiz, Braier, Exxon Valdez* and *Berman* will continue to haunt us and the world's marine resources.

WHO PAYS WHEN OIL SPILLS?
The International Scheme

SACRIFICING ENVIRONMENT ON THE ALTAR OF BUSINESS

For as long as ships have sailed the seas, man has done so with impunity. Right of passage through territorial waters has guaranteed freedom of navigation across the oceans of the world. For too many years the environment has been sacrificed on the altar of business. Only since 1967 and the *Torrey Canyon* oil spill has there been a change.

A threatened or actual marine pollution incident triggers set obligations of the shipowner or cargo owner: first to the ship, the crew and the cargo; next to the affected states (nations or countries) or territories; then to the insurer to minimize the insurer's liability for cleanup and salvage costs under the *sue and labor* clause of the policy; and lastly to a contracted salvager under *The International Convention on Salvage, 1989* (when that treaty goes into effect) if the vessel needs to be removed.

Supplementing these duties to act are compensation requirements imposed on the shipowner by international regimes and recent United States laws. These plans charge the vessel owners with the cost of cleanup and damage to people, property and the environment. After *Torrey Canyon*, four international regimes came into being, two voluntary schemes regulated by the insurance industry and two conventions supported by countries and nations (states) signatories to these international treaties. These plans address the question of who will pay the costs in a threatened or actual release involving pollution of the seas. With the 1992 Protocol amendments to one major international treaty, the range of coverage under international treaties is now about $210 million for each incident.

The U.S. is not a signatory to the conventions, nor is it a party to the voluntary insurance regimes. We have taken the position under OPA 90 that in certain cases there is *no limit* to what the polluter will be required to pay. Our unwillingness to join in the international strategies has earned us a *bad reputation*. We are regarded by some in the international and business communities as being "out of step," "unreasonable," using a "bargain basement approach," and much worse.

Under U.S. laws since the early 1970s, shipowners have been required to carry certain minimum insurance to pay costs in the event of an oil discharge. Recently Congress raised that amount significantly. Each vessel trading in U.S. waters must show proof of insurance, evidenced by a Certificate of Financial Responsibility (COFR). The USCG is charged with issuing these certificates and enforcing their use. When the financial responsibility coverage of a Responsible Party (RP) is insufficient to meet the costs of cleanup and restoration of the damaged environment, the U.S. Oil Spill Liability Trust Fund kicks in for the balance of damages up to $1 billion per incident. Of this total, not more than $500 million can be paid for natural resource damages per incident. (While these funds are available, the most recent opinion issued by the U.S. Comptroller General does not allow compensation for these natural resource damages without specific appropriation approval from Congress. This opinion is under attack, but remains in effect and may delay recovery for damage to natural resources.)

In the U.S. these limits of liability are not the maximum penalties. They can be broken under OPA 90. The shipowner can be found liable for every penny of damage over and above insurance limits, if one of the following occurs:

- The incident was caused by willful misconduct, gross negligence or violation of any federal safety, construction or operating regulation; or

- The polluter failed to notify proper authorities of the incident; or

- The polluter failed to reasonably cooperate with the OSC in cleaning up the spill or to comply with an order under the Clean Water or Intervention Act.[1]

In addition, polluters may also pay civil penalties and fines and serve time in jail. These civil penalties and fines are not subject to OPA 90 limits.

The insurance-driven shipping industry worldwide bitterly opposes the U.S. approach. We are in a fight with big business. Whether our collective intent to continue making the polluter pay survives the present political climate in Congress remains the question.

INTERNATIONAL REGIMES

Two international conventions known as CLC and the Fund Convention pay out damages in the event of an oil spill.

The International Convention on Civil Liability for Oil Pollution Damage, 1969 (CLC) was the first of the two conventions to make an owner of a tanker liable for pollution damage, regardless of fault. While the conventions speak of *strict liability* (liability without regard to fault), responsibility for damages under these schemes is subject to severe limitations:

- Not all oil discharges are covered. Exposure to damages applies only to discharges of persistent oil into the marine environment. Persistent oil is crude, fuel, heavy diesel, lubricating or whale oil. There is no compensation for a spill of *nonpersistent* oil, such as gasoline, kerosene and light diesel oil.
- The geographic scope of coverage is limited to the area of damage suffered in the territory or the exclusive economic zone of the party to the convention.
- A shipowner can assert defenses to liability such as an act of war, an act or omission of a third party, or the negligence of some governmental or other authority in failing to maintain lights or navigational equipment.
- There is a cap to the amount that the shipowner pays based on vessel size and a per ton amount. The vessel owner is responsible up to the stated limits, unless the incident resulted from actual, personal fault of the shipowner (1969 CLC) or was caused by a "personal act or omission, committed with the intent to cause such damage, or recklessly and with knowledge that such damage would probably result." (1992 Protocols.)

The 1992 Protocols amend the 1969 CLC, effective May 1996, and provide the following:

- Extend coverage to unladen ships as well as ships carrying oil in bulk.
- Allow for compensation in the case of a threat of discharge and for preventive measures taken when there is a *grave and imminent* danger of pollution damage and the responsive actions taken are likely to be successful.
- Owners of tankers carrying more than 2,000 tonnes of persistent oil must have sufficient insurance to cover this amount of liability and must display proof in the form of an on board certificate.

The second major convention is the International Convention on the Establishment of the International Fund for Compensation for Oil Pollution Damage, 1971 (the Fund Convention). This convention authorizes use of the International Oil Pollution Compensation Fund (IOPC Fund), established by the international community in 1971 to administer claims and pay out damages.

The Fund Convention compensates claimants when CLC funding is insufficient to cover spill damages. Obtaining compensation is not easy.

A shipowner has no payment obligation if the discharge is caused by natural phenomenon, if the damage was wholly the fault of a third party, or if the governmental interference exception applies. In all other cases, the IOPC Fund compensates those injured up to the stated maximum of about $93 million. The 1992 Protocols will increase the aggregate amount payable under both conventions to $210 million for claimants in those countries which have ratified these latest Protocols.

How is the IOPC Fund financed? Any person or entity whose country (called "States" in the Fund Convention) is a party to the Fund Convention and who receives more than 150,000 tonnes of crude and heavy fuel oil in one calendar year must pay an *initial* contribution as a member, an *annual* contribution based upon oil received and, if necessary, a *general fund* contribution if too many spills happen in a given year and necessitate a call for additional capital. The bitter irony in this funding scheme is that oil-rich exporters, such as many of the Arab countries, do not pay a cent into the Fund, although they benefit the most from transporting oil.[2]

VOLUNTARY AGREEMENTS – TOVALOP

While the conventions were being negotiated, industry set up two additional compensation schemes. TOVALOP is an agreement entered into by *tanker owners* and bareboat charterers to assume obligations not otherwise required by law. Ninety-seven percent of the world's tanker tonnage is party to TOVALOP, which is also covered by the Shipowner's Prevention and Indemnity Club (P & I Club). Supposedly there is *strict liability* with only limited defenses. But the same convention limitations apply.

TOVALOP does not kick in if the incident is fully covered under the CLC, and applies only if the incident was caused by a tanker owned by a party to TOVALOP. TOVALOP coverage is available in a *pure threat situation* where there is real danger of a discharge. There have been several supplements to TOVALOP, so the current total coverage for tankers up to 5,000 gross tons is $83.6 million per incident.

Each shipowner maintains insurance with one of the 18 P & I Clubs worldwide. P & I policies provide insurance for pollution damage, cargo damage, death and injury to crew – but not for ship, hull or machinery damage. The P & I Club pays up to its limit, depending on which one of their members caused the incident or where the discharge occurred. After the first $20 million is paid out, then each of the clubs pools resources to come up with another $20 million. The P & I Clubs resort to buying *reinsurance* on the commercial market if necessary. In the final analysis, the consumer pays. Companies generally pass the cost of insurance on to consumers in the form of a higher per gallon gas charge.[3]

PAYOR OF LAST RESORT – CRISTAL

The second voluntary agreement sponsored by industry is CRISTAL, which involves the *cargo owners* in the reparations schemes. Under this agreement, cargo owners provide supplementary compensation for claims under CRISTAL only when the limits of TOVALOP are exceeded.

All four schemes work together. Where an IOPC Fund member is involved, CRISTAL kicks in only after the CLC and Fund Convention aggregates and TOVALOP Supplement have been reached, and all reasonable steps have been taken to obtain compensation from other sources (such as the parties at fault). CRISTAL is the *payor of last resort*. The total compensation available under the two funds is $46 million for tankers up to 5,000 gross tonnes, and up to $186 million for ships exceeding 140,000 gross tonnes.

In November 1995 ITOPF took a dramatic step and voted to terminate the voluntary schemes effective February 1997. With the 1992 Protocols in force, ITOPF believes that claimants have access to at least twice the amount available under the voluntary schemes, rendering them obsolete. By withdrawing the voluntary schemes, ITOPF hopes that more countries will be forced to ratify the conventions. The ultimate impact of the loss of TOVALOP and CRISTAL on the polluter and claimant is unknown.[4] With so much insurance coverage, the obvious question is: why isn't the U.S. a party to these conventions and voluntary agreements? The answer cuts to the heart of the debate, and underlines the basic philosophical difference between the U.S. and the rest of the world regarding oil spill liability.

POLLUTION DAMAGE COVERAGE

Nowhere does the contrast between an *efficient* response and an *effective* response appear more clear than in the area of what type of damage is covered by the four international regimes. From the U.S. point of view, the goal of these coverages appears to be *exclusion* as much as possible. To the international community, the operative word characterizing the U.S. scheme is *unreasonable*.

The 1992 Protocols add to the definition of *pollution damage*, "loss or damage caused outside the ship carrying oil by contamination resulting from the escape or discharge of oil from the ship, wherever such escape or discharge may occur" a phrase providing for, with regard to environmental damage, the costs incurred for reasonable measures to restore the contaminated environment.[5] This does *not* mean that all damages are covered.

Fund and CLC technical advisors generally recognize two types of preventive measures: 1) the removal of oil from a damaged tanker; and 2) cleanup measures after removal, including disposal of recovered oil and oily debris.[6] That may sound fairly comprehensive. But a closer look at the definitive *Berman* cleanup reveals the marked differences between the U.S. and international schemes.

ITOPF's technical staff often work hand-in-hand with the IOPC Fund to observe and advise at spill sites and later resolve questions of coverage. In January 1994 ITOPF issued a series of working group notes intended to clarify which claims the federation staff believes are admissible. The Seventh Intersessional Working Group of the International Oil Pollution Compensation Fund appeared to adopt these recommended notes as announced in their June 20, 1994 report. The ITOPF Working Group notes serve as a guideline to classify damages the voluntary schemes and conventions will recognize.

"HOOKED ON A RUNAWAY"

Reasonableness in terms of preventive measures yields this definition of what will be approved and what will not:

> *"Whilst preventive measures are subject to the test of 'reasonableness,' the term is undefined in the Conventions. However, it is generally interpreted to mean that the measures taken or equipment used in response to an incident were, on the basis of a technical appraisal at the time the decision was taken, likely to have been successful in minimizing pollution damage. As a general rule the measures should be expected to enhance the natural process of oil removal. The fact that the response measures turned out to be ineffective or the decision was shown to be incorrect with the benefit of hindsight are not reasons in themselves for disallowing a claim for the costs involved. A claim may be rejected, however, if it was known that the measures would be ineffective but they were instigated simply because, for example, it was considered necessary 'to be seen to be doing something'. On this basis, measures taken purely for public relations would generally be considered unreasonable."[7]*

The emphasis in this ITOPF directive is on *effectiveness, reasonableness, technical appraisal, minimizing pollution damage* and *enhancing the natural process*. How does this focus translate to cleanup in the field? Responses are based on economy of scale, from the size of the management team to the size of the work force. The use of dispersants is approved in the right circumstances to accelerate weathering of oil. Recovery at sea receives little support as an effective means of capturing oil, especially after the initial days of the spill. As to shoreline cleanups, the general approach is that "most coastal oil spills will result in pollution of shorelines despite efforts to combat the oil at sea and protect the shoreline."[8]

Once the bulk oil is removed, secondary and further cleaning does not appear to be appropriate in the majority of cases. Disposal of debris is subjected to the most *cost-effective* alternative, as determined by fund staff. Wildlife rehabilitation is seen as rarely effective and "difficult and slow." Commercial cleanup companies may act at their peril if later on their cleanup measures were deemed not to be effective by ITOPF or Fund staff. Without contracts, they may not be paid for their work. Costs for personnel are to be kept to a minimum, only incurred so long as needed, and based on market rates.

These strictures would have eliminated all but bare minimum cleanup in *Berman*. Instead, Capt. Ross's approach was to hit the problem frontally, proceeding with maximum effort to:

- Recover as much oil as possible on the water.
- Recover gross contamination and clean beaches until they met *how clean is clean* guidelines set by the FOSC, NOAA and various support coordinators for the *Berman* cleanup.
- Retrieve submerged oil.
- Protect sensitive areas.
- Rehabilitate as much affected wildlife as possible.
- Haul off as much oiled debris as could be recovered.
- Return cleaned sand back to the beaches.

Safety, not money, was the primary consideration.

In *Berman* the U.S. approach translated into a harsh fact. That the USCG spent some $82 million ($5 million of which was reimbursed) on cleanup provoked the response from others in the international field that the U.S. was "hooked on a runaway."[9] To detractors, we have caught our foot in the stirrup of high-tech machinery and recovery devices. They view our dependence on technology as an accelerated departure from a common sense approach. We are simply pouring money down the drain in considering this approach as leading to a successful cleanup. The international community believes such funds would be better spent on providing basic needs like food, housing and clothing for the residents of affected communities.[10]

Joe Nichols of ITOPF shares this view. His critique of the *Berman* spill response underlines a basic philosophical difference. In his opinion almost the entire response should have been conducted differently:

- Too much emphasis was placed on skimming oil from the water.
- Certain beaches should have been sacrificed.

- Too much duplicate cleaning of sand took place.
- Too many men and machines were employed in the task.

What the end result in cleanup would have been under these limitations is anyone's guess.

According to the Preassessment Screen Document, the *Berman* spill was potentially disastrous. About 169 miles of coastal shoreline, some 292 acres of seagrass, about 5,120 acres of wetlands, 1,100 square miles of surface water, and 1,900 acres of bottom sediments were exposed to the heavy No. 6 fuel oil release. Because of the high resistance of this viscous oil to degradation by molecules and weathering, long-term environmental effects are expected. However, based on the preassessment damage report, Capt. Ross can only be praised for his efficient and complete response. Money spent on the front end in a quick and full cleanup resulted in much less damage to the environment and health of the people and creatures of the area. According to the preassessment report:

> *"Emergency restoration activities are undertaken when no action would result in a measurable, avoidable increase in natural resource injuries or losses of services. Removal and cleanup operations have eliminated any currently apparent need for emergency restoration, with the possible exception of residual impacts to seagrass resources ... The response actions carried out by the Federal On-Scene Coordinator coordinated with the Coast Guard and other commonwealth and federal agencies have contained elements which have attempted to prevent further injuries."*[11]

In other words, Capt. Ross's strategies worked – to a point. With the exception of seagrass beds, other emergency activity was unnecessary. However, even the best of responses cannot sufficiently restore or compensate for lost or damaged natural resources or services. Restoration plans are needed to address unavoidable oil impacts. In *Berman,* most environmental injuries occurred at the time the oil contacted the resources. Cleanup and containment did prevent a much greater long-term impact and the need for emergency action after the initial three months to salvage ecosystems in the path of *Berman* oil. There is a strong opinion by the majority of coordinators involved in the response that the preassessment screen would have been significantly different in terms of damages to natural resources if the ITOPF approach ruled the day.

Scientific evidence exists to confirm that thorough cleanup efforts *do* aid in restoring balance to a natural resource or habitat damaged by an oil spill. Dr. Mark Miller's scientific work with marine creatures in Punta

BURSATELLA LEACHI

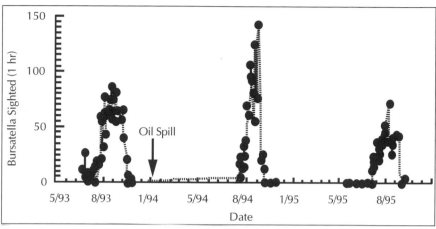

(Source: Reprinted with permission from Dr. Mark Miller, Institute of Neurobiology, University of Puerto Rico.)

Escambrón, the site of the *Berman* grounding in January 1994, supports the effectiveness of Capt. Ross' hard-hitting cleanup approach. Dr. Miller studies the *Bursatella leachi*, a bottom-dwelling mollusk that inhabits a protected cove near Punta Escambrón between July and November annually. According to his pre-spill baseline data and post-*Berman* research, the cleanup proved so effective that the number of specimens observed in July 1994 and July 1995 exceeded that of 1993. Dr. Miller cautions that interpretation of such data is not always straightforward. For example, a predator of the *Bursatella* may be severely affected by the oil spilled, giving rise to an increase in the numbers of this species.[12]

NO BETTER OFF

Under the international approach, *cost* is supreme. Only direct, observable losses are compensated. The IOPC Fund reimburses a claimant for costs of cleaning polluted property, for example, fishing gear, boats, yachts, beaches, piers and embankments (after deduction for normal wear and tear), or for replacing gear which cannot be cleaned. The overriding principle is that *"the claimant should be no better or worse than if the spill had not occurred."* This translates into a strict process of documenting what was lost, then deducting depreciation before payment of any property damage.

Economic loss caused by the spill is even more severely limited and compensation is paid only to those who depend directly on earning a living from coastal or sea-related activities, such as fishermen or hoteliers and restaurateurs at seaside resorts. The fact that a spill occurred in a heavily used recreational area does not result in automatic compensation. Payment is limited by several conditions:

- Proof that economic loss was caused by the oil *contamination or preventive* measures.
- Evidence of real loss.
- Claimant must show he took reasonable measures to mitigate his damages or that such actions were impossible.
- Money recovered elsewhere during the spill is deducted from overall compensation.

For those individuals even slightly removed from the spill, their losses receive less credence. The rule states, "A claimant more remotely linked both in terms of business and geographical location would incur little or no loss ..."[13]

These conditions translate into *limiting* coverage. For example, the principle governing fish catch is that an oil spill is a "short-term interruption and unlikely to affect the overall fishing quota." For cultivated marine organisms or commercially grown stock in the *mariculture* industry, ITOPF considers that lost growth potential is often recovered in subsequent weight gain. Future losses in general are viewed as "speculative, since it is rarely possible to demonstrate that such losses will definitely occur," with limited exception.[14]

Losses from tourism are defined in similar terms: "There is little evidence for longer term local impacts on the tourist trade from spills. Once beaches are clean, normal trade should resume ..."[15] One wonders under the international scheme how the Puerto Rico Condado area businesses would be treated by ITOPF, considering the spill shut down their beaches for between six weeks and three months at the height of the tourist season and the adverse national and international publicity discouraged tourists and travel agents.

IMPAIRED ENVIRONMENTS

The area most affected by the cost-cutting knife of international regimes is *environmental damage*. Initially only *quantifiable economic loss* was recognized by the regimes. The Protocols add *impairment of the environment* as a subject of restitution. The IOPC Fund will pay "costs of reasonable measures of reinstatement actually undertaken or to be undertaken." How does this equate to real restoration or repair of damaged ecosystems?

Again, certain principles put blinders on the eyes of the Fund compensators. There is *no compensation for any environmental resources which do not have a commercially exploitable market value.* For example, the public's lost use of a beach, a coral reef or any such natural resource is excluded entirely. The Fund is loud and clear about which damages are compensable and which are regarded as highly theoretical and speculative:

> *"The assessment of compensation to be paid by the International Oil Pollution Compensation Fund is not to be made on the basis of an abstract quantification of damage calculated in accordance with theoretical models."*[16]

What does this mean exactly?

The marine environment has a value to the average person far and above what it provides in the way of livelihood for fishermen or fisheries. We assess a given natural resource or ecosystem based on intangible qualities. For example, the sheer existence of a coral reef is valuable to many. This intangible quality has been called *non-use or passive use value.* Other terms used to describe this quality are option, existence, bequest or heritage value.

The Fund denies compensation for such values. The rationale for refusing to pay for non-use loss of these natural resources is that one cannot put a commercial price tag on them. Therefore their value is "highly theoretical and speculative." Oil spills interfere with natural resource amenities for a "short time" only.

"The marine environment is highly resilient to short-term changes, and numerous studies have demonstrated that a major oil spill will not cause *permanent* effects, except in truly exceptional circumstances," notes ITOPF.[17] Fund supporters express this opinion in spite of the numerous studies and conclusions drawn by scientists and environmentalists to the contrary: oil spill damage to many natural resources is clearly long-term and in some cases permanent. (See Section II, Chapters 12 and 13.)

The international schemes cover very little environmental damage – only the costs for reasonable measures actually taken to reinstate the affected environment, assuming that such restoration is even possible. Once *cleanup* is completed, the question of *restoration* of an affected area or affected wildlife and plants arises. The Fund starts from the position that "the potential for natural recovery is great and that man is severely limited in the extent to which he can take restoration measures which will improve upon the natural processes."[18]

According to Fund advisors, Mother Nature will restore most, if not all, of the environment in time. Even when destruction is widespread and cannot be ignored, replacement of an affected site by rehabilitation of another site near the spill or within the region of the spill is not feasible, practical or reasonable. This view means the polluter has no responsibility to take any corrective action if seagrass beds destroyed by an oil spill in a discharge zone will *not* regenerate over time. Even where money could be used to allow for cleanup of waste water being discharged directly onto another adjacent seagrass bed, the Fund sees no reason for the polluter to do anything to benefit the ecosystem on a regional or alternative basis. The fact that the spiller displaced thousands of marine organisms by destroying their habitat has nothing to do with providing an alternate place for those organisms to relocate. Such a program would be considered unreasonable.

In summary, the Fund will pay for compensation measures for reinstatement of the environment only if the following criteria are met:

- The costs of the measures are reasonable.
- The costs are not disproportionate to the results achieved or reasonably expected to be achieved.
- The measures are appropriate and offer a reasonable prospect of success.[19]

QUICK SETTLEMENT

There are some clear advantages to the four international regimes. ITOPF and other Fund technical teams are available to help assess and monitor all spills. The schemes have international support. As of the *Berman* spill in 1994, 87 nations were party to the CLC and 60 to the Fund Convention. Losses of a purely economic nature can be processed and settled relatively quickly. The director of the Fund can settle claims directly with individuals or small businesses up to an amount of $1 million from Fund sources without executive committee approval and with minimal delay.[20]

WHO PAYS WHEN OIL SPILLS?
The U.S. Scheme

THE POLLUTER PAYS

The U.S. scheme has caused a furor among international insurers and shipowners. After *Exxon Valdez*, Congress took a significant step intended to motivate oil companies and the shipping industry to reduce the probability of continued oil spills and passed OPA 90 into law.

The intent behind this legislation was to create strong economic and other incentives for the potential *Responsible Party* (RP) to act more carefully. OPA 90 holds an RP strictly liable for removal costs and a wide range of other damages caused by oil, including damages to the environment. The RP, up to the limits of that party's liability, is obligated to pay all damages. In certain instances the RP may be criminally liable and may pay considerable civil penalties as well, without the liability limitations imposed by OPA 90. Additionally, U.S. law provides for substantial penalties of up to $25,000 a day for any violation of OPA 90, and for possible forfeiture of the vessel. While this scheme sounds similar to the international regimes, the two systems are oceans apart.

OPA 90 is far-reaching legislation that accomplishes many fundamental goals:

- Setting new requirements for vessel construction.
- Controlling crew licensing and manning.
- Mandating contingency planning.
- Enhancing federal response capability.
- Broadening enforcement authority.
- Creating a new research and development program.
- Addressing risk exposure and financial responsibility for polluters.

Unlike the international regimes, which include only *persistent* oil, the term *oil* under the U.S. scheme is broadly defined to mean most petroleum products except hazardous substances already covered by CERCLA, such as benzene, xylene and toluene.

Under OPA 90, the RP can be the vessel owner and/or operator. The regimes provide only for owner responsibility. Ships covered under OPA 90 include any vessel greater than 300 gross tons, and lightering vessels used to offload oil from one ship to another. (OPA 90 covers land-based oil facilities as well.) The international conventions apply to ships carrying more than 2,000 tons of oil in *bulk as cargo*, or an unladen ship carrying residual oil.

The geographic scope of OPA 90 is not limited to the pollution damage caused by oil escape or discharge on the territory or the exclusive economic zone of a party, as in the international schemes. OPA 90's geographic range includes U.S. ocean, coastal and inland waters, protecting all traditional navigable U.S. waters and surface waters. Threatened discharges are part of the scheme as well.

OPA 90 limits defenses to a claim for damages to an act of God, war, or act or omission of a third party. The international schemes establish additional defenses such as negligence of a government in maintaining lights or other navigational aids.

OPA 90 and the international regimes depart significantly in terms of what type of damages are *recoverable*. OPA 90 covers the following:

- *Real or Personal Property Damage* for injury to or economic loss from destruction of property.
- *Loss of Profits or Earning Capacity* for damages equal to the loss of profits or income due to destruction or loss of property, including damage to natural resources.
- *Removal Costs*, the costs to remove the discharge, including those to prevent, minimize or mitigate a substantial threat.
- *Loss of Subsistence*, the loss of use of natural resources until they recover to pre-spill conditions.
- *Loss of Government Revenues*, referring to net loss by a governmental entity of taxes, royalties, rents or fees from damaged property.
- *Cost of Increased Public Services*, the additional costs incurred by a governmental entity for providing public services such as fire, safety and health protection.
- *Natural Resources*, damages for injury to, destruction of, loss of, or loss of use of natural resources, including the costs to assess these damages.[1]

THE LIABILITY LOOPHOLE

RPs face limits to their financial liability:

- For tankers greater than 3,000 gross tons, liability is the greater of $1,200 per gross ton or $10 million.
- Tankers less than or equal to 3,000 gross tons are liable for the greater of $1,200 per gross ton or $2 million.
- For non-tankers, liability is the greater of $600 per gross ton or $500,000.

These liability limits do not seem oppressive when compared with the international compensation plan. However, the crucial difference between the U.S. and global regimes is that *limits can be broken* in the U.S. This is where the insurance companies turn apoplectic. Under the international schemes, the only way to breach policy liability limits is if the owner is actually at fault. The 1992 Protocols classify pollution damage as that which results from the owner's "personal act or omission, committed with the intent to cause such damage, or recklessly with knowledge that such damage would probably result." Any good trial lawyer will agree that *intentional acts* can be extremely difficult to prove.

The liability limits are more easily breached under OPA 90 by a finding that the incident was caused by "willful misconduct, gross negligence, or violation of federal safety, construction or operating regulation." Additionally, the failure of the RP to report the incident, or *to provide reasonable cooperation to officials in removal activities*, or to comply with an order under Section 311(c) or (e) to the Clean Water Act or Intervention Act, voids the liability limits.[2]

Berman displayed numerous violations of federal safety, construction and operating regulations. Gross negligence and willful misconduct existed. Perhaps no one intended for the incident to happen, but the actions of the principles and their agents guaranteed an oil discharge. The Berman/Frank companies will be unable to assert limits to their liability because of their failure to comply with safety and operating regulations, their gross negligence in failing to repair the frayed towline, and in numerous other inexcusable acts of the company, its officers and agents, discussed at length in earlier chapters.

Because of these exceptions to liability limits, under OPA 90 the U.S. can hold a shipowner responsible for any damage caused by a spill, *without limitation*. The probability of breaching limits has the shipping industry and international communities in an uproar. How do they view this U.S. provision? According to ITOPF spokesman Joe Nichols:

> "Under the International Compensation schemes, costs have to stand the test of reasonableness. Under OPA 90 there is no reasonableness. A lot of people simply show up with their own agendas. Add to this your liability scheme, and the U.S. is at best a lost cause.

*If shipowners could depend upon a limit to their liability of the statu-
tory amount, $1,200 per gross ton, that would be reasonable. But
shippers to the U.S. know that claims can go way beyond because of
excessive cleanup costs, class actions by fishermen, and natural
resource damages.*

*"Today any shipowner dealing with the U.S. can get $500 mil-
lion maximum insurance. After that, he knows that he stands the
risk of paying directly out of his own pocket. This is so onerous that
any company of decent standing will simply not bother trading with
the U.S. What worries me the most is that the U.S.'s bad habits
might be spreading, for example to Canada. The United States has
opted for the 'easy way out.'"[3]*

John Schrinner, implementation officer of IMO, offers this view:

*"Since the U.S. is not a party to the civil liability conventions and
there is a possibility of unlimited liability, companies are very wary of
what ships to send to the U.S. With a new vessel only ten to 12 years
old, with at least eight good years left, there's a lot of value to that ship,
maybe $30 million to $40 million. The shipowner runs the risk of losing his entire ship unless he's prepared to dot all i's and cross all t's.*

*"The end result is that company owners may choose to send their
cheapest ships or to charter vessels from Flag States with countries
having poor standards. With no business assets or holdings in the
U.S., the only recourse will be to a junker with a value tops of $5 mil-
lion to $10 million. This translates into substandard shipping in
U.S. waters and defeats OPA 90. I call the U.S. scheme a 'bargain
basement approach.'"[4]*

Those close to the U.S. system agree in part, but still support the concept that the polluter pays:

*"The reaction on the part of major oil companies is that they are
being more conscious and taking more aggressive positions to prevent pollution. While you might be skeptical about their motivation,
that doesn't matter. Hitting these majors hard and heavy works. The
bottom line is that the polluter pays. For many foreign states and the
nickel and dime operators, the OPA system means they can't comply.
They don't have the financial wherewithal. Eventually this will
shrink their spheres of operation."[5]*

FUNDING SOURCES: OSLTF

A major source of funds to pay for the cost of cleanup and damages comes from the Oil Spill Liability Trust Fund (OSLTF), a $1 billion fund established initially under the IRS Code in 1986. In the wake of *Exxon Valdez*, Congress passed legislation under OPA 90 allowing access to these funds. OSLTF was set up to accomplish specific tasks:

- To provide money for removal actions.
- To pay costs for assessment and restoration of damaged natural resources (subject to a recent U.S. Comptroller general opinion regarding Congressional appropriation).
- To provide compensation to claimants.
- To pursue recovery from RPs for removal costs and damages paid out by the fund.[6]

In theory, the OSLTF is like the international funding agency, the IOPC Fund. The National Pollution Fund Center administers the OSLTF. The claims system works this way: the claimant first presents his damages to the RP or the RP's insurer, if one is identified. The RP has 90 days to respond. If there is no response, or if the RP elects not to pay the claim, then the NPFC becomes involved, receiving and processing the claim.

Recently, on October 30, 1995, the Comptroller General of the U.S. issued an opinion that has modified the OSLTF's procedure for processing claims. The ruling states that trustees must obtain an appropriation before putting in a Fund claim (based on fiscal law principles that override what appears to be the plain language of OPA 90.) Based on this opinion, trustees can only obtain money from the Fund (except for the funds needed to "initiate" an assessment) through Congressional appropriations. This written opinion applies to all trustees, federal, state and tribal. The state of New York has challenged the ruling in federal court. Sen. Chafee has introduced a bill in the U.S. Senate, a portion of which seeks to override this ruling. While the ruling stands, claim processing will be less expeditious than before.[7]

Recognized claimants are damaged individuals, the U.S. government, states, territories and commonwealths of the U.S., foreign claimants, and trustees whose natural resources are damaged by the spill. The trustees are the only persons or entities able to seek damages for natural resources. Federal trustees recognized by OPA 90 include the National Oceanic and Atmospheric Administration (NOAA) for the Dept. of Commerce, and the departments of the Interior, Agriculture, Energy and Defense. U.S. state governments, tribal nations and foreign trustees can also file claims for natural resource damage.

The OSLTF is funded ultimately by you and me. The consumer bears the cost which the oil companies pass through in additional charges at the gas pump. Initially there is a five-cent-per-barrel tax on domestically produced or imported oil. In this manner the oil industry fronts the fund. If the OSLTF falls below $1 billion, then the tax is reinstated. For example, in July 1993 the OSLTF faced depletion and the tax was reinstated to build up the pool for damage recovery. This tax on companies provides a significant part of the pool, 83 percent of all revenues in 1993. Other sources for the OSLTF are interest on U.S. Treasury investments, civil penalties levied against RPs, and costs recovered from RPs by the NPFC.

While the international community might consider the amounts paid under the OSLTF to be huge, there are limits. States can only apply for up to $250,000 for removal costs from the Emergency Fund. There is a cap of $1 billion in any one incident from the General Fund, which includes a maximum compensation payable for natural resource claims of $500,000 an incident. (Again, until the Comptroller General's opinion is settled by litigation, these amounts are subject to Congressional appropriation and cap limits become moot.)[8]

$300 FOR COMPENSATION

How quickly are claimants paid? If all the paperwork is in order, claims can be settled within 90 days. In *Berman*, claimants accepted settlement offers in 99 percent of the 873 claims filed.[9] But organizing all the information necessary to document claims is another matter.

Marisol Luna, the wife of a longtime Puerto Rican fisherman, described the problems many Puerto Rican fishermen had in proving their *Berman* spill losses and claiming compensation from the OSLTF.

Luna's husband is a commercial fisherman who dives to catch his fish and lobster. In 1994, he was 35 years old and had fished for 26 years. The *Berman* spill caused economic hardship for Luna and other fishermen in the Isla de Cabras area of Puerto Rico. Fish and lobster production dropped from between 30 pounds to 50 pounds of lobster a day before the spill to about one to three lobsters daily after. Small fish in the area died en masse. Marisol claims that months later the catch still lagged far behind traditional figures. During the Easter Holy Week 1994, the catch ran about 125 pounds of fish, compared with 900 pounds for the same period in 1993. Marisol and others believe the fish they catch still smell of oil. She recalls cleaning oil from her husband's hair with degreaser nine months after the spill.

"Surely these fishermen should be compensated," one might argue. Yet one year after the spill, in Marisol Luna's association of some 200 fishermen, the most any fisherman had received was $300 each from the Emergency Fund.

The problem lies with the evidence supporting the previous years' catch. Usually the appropriate agency in the Puerto Rican government would have paperwork showing earlier years' fish catch. However, many of these old records were tossed or misfiled and the government cannot retrieve the information. When this situation became apparent, NPFC adjusters agreed to accept reports directly from the fishermen. Even this solution failed. Many of the fishermen involved cannot read or write. They can judge the size of a fish almost to the ounce and set a price accordingly, but maintaining a daily account of how many pounds of fish they catch is beyond the routine of many of the fishermen.

Even those who could write did not keep a neat set of ledgers. Many simply scribbled the amount of their catch on the walls of the Fishermen's Association "house." A fishermen might note on the walls, "August 5, 12 pounds of grouper, 11 pounds of red fish." Marisol Luna has taken on the unenviable task of compiling in writing whatever records she can find for each fisherman. One answer may be to send the walls of the house into the NPFC as proof![10]

Assuming a claimant can clear these hurdles, once a claim is approved (and once a Congressional appropriation has been made), the claimant signs a release and receives a check. The check generally represents full compensation, although in a hardship case, the fund can make an interim payment and later settle once all evidence is provided. This is the procedure used with the fisherman of Isla de Cabras. Their interim payment was about $300 per person. Once paid, the Dept. of Justice, on behalf of the OSLTF, is charged with suing the RP to recover all removal costs and damages.

The claims covered under OPA 90 are more broad and comprehensive than those recognized under the international regimes. While there are some drawbacks, OPA 90 claims processing seems to run as smoothly as it does under the international schemes. Strict liability under OPA 90 means the polluter will pay more than finite economic damage. Liability limits can and will be breached where *Berman*-type violations of safety and operating regulations or gross negligence exist.

The *Berman* spill demonstrates an even greater impact of OPA 90 in terms of who pays: the polluter faces criminal felony conviction. In May 1996, a jury in the U.S. District Court in San Juan, P.R. convicted three corporations: the Bunker Group-Puerto Rico, the Bunker Group, Inc., and New England Marine Services; and convicted one individual, Pedro Rivera, the general manager of Bunker Group, on felony charges stemming from the *Berman* grounding.

Rivera faces up to five years in jail. The corporations may be liable for as much as *twice* the gross financial damages and response costs which are in excess of $87 million. The tug captain, Roy McMichael, Jr., and First

Mate Victor Martinez pled guilty to criminal negligence charges in the *Berman* incident. Their misdemeanors carry up to a one-year jail sentence and substantial fines, as well as loss of license by the USCG governing board.[11]

In the U.S. under OPA 90, *the polluter pays* can result in severe economic consequences for irresponsible oil shippers.

THE COST OF RECOVERY

*Natural Resource Damage Assessment (NRDA)
and Certificates of Financial Responsibility (COFRs)*

The insurance industry has decided to draw the line against OPA 90 regime requirements in two areas: recovery for natural resource damage, and the increased requirements of the new Certificate of Financial Responsibility (COFR). The international community and many in industry claim that the OPA 90 scheme of recovery for natural resource damages is "junk science."

THE STRAW BREAKING THE CAMEL'S BACK
Joe Nichols, ITOPF:

> *"'The straw that's breaking the camel's back' is the U.S. reliance on NRDA. This is junk science at its worst. The U.S. has opted for the simplest method consistent with getting the biggest pot of gold. The formula approach yields a half million dollar recovery for one gallon of crude spilled in San Francisco Bay in the summer. What's even more ridiculous is that the money recovered from a spill isn't even being used to clean up that spill. Instead, recoveries are being pooled to clean up other parts of the region having nothing to do with the spill. To the international community, the U.S. NRDA scheme is immoral."[1]*

What is the NRDA argument all about? Is the U.S. guilty of using "junk science" to penalize the polluter? Or are we on the right track?

Recovery of damages for natural resources is a creature of law. Until the National Oceanic and Atmospheric Administration (NOAA) promulgated its final regulation concerning NRDA in January 1996, there were two systems controlling damage assessments for oil spills. One was regulated by the Dept. of Interior under the Comprehensive Environmental Response, Compensation and Liability Act of 1980 (CERCLA; 42 U.S.C. Section 9601 *et seq.*) and the Federal Water Pollution Control Act (Clean Water Act; 32 U.S.C. Section 1321 *et seq.*) The other was regulated by NOAA under the Oil Pollution Act of 1990 (OPA 90) (Public Law 101-380

(1990). These three pieces of legislation created responsibility not only for the cost of cleaning up releases, but also for monetary damages for injury to natural resources caused by oil discharges.

On January 5, 1996, NOAA issued its Final Regulation on NRDA, contained in 61 Fed. Reg. 440 as codified in 15 C.F.R. Part 990. Since January 1996, NOAA regulations set the stage for valuing damages caused by oil releases for restoration purposes and determining the loss of these services until their restoration to prespill condition.

DOI regulations apply only to hazardous substance releases, except in three limited instances:

- If a damage assessment for oil spilled was underway in January 1996, then that assessment will proceed to conclusion following DOI rules or NOAA regulations.

- NOAA allows use of *models,* referred to as the *Type A model,* in certain cases to value damage caused by a spill. Trustees can look to DOI's Natural Resource Damage Assessment Model for the Coastal and Marine Environment, and to the model for the Great Lakes Environments for purposes of valuation (found in the May 1996 Final Rule for Type A assessments, Federal Register Vol. 61, No. 89).

- Industry challenged NOAA's NRDA regulation. Petitioners in that lawsuit argued that a case pending before the same court in which DOI's NRDA rule was at issue contained substantially similar points, and that the outcome of the DOI suit should influence resolution of issues concerning the validity of the NOAA regulation (more on this subject later in this chapter).

Except in these special instances, the prevailing rule for reference on how the U.S. assesses natural resource damage caused by an oil spill is the January 5, 1996 NOAA final rule.

Congress charged the Dept. of Commerce through NOAA, as the principal federal trustee for marine environmental natural resources (along with states and tribal nations as other concerned trustees), to recover monetary damages from Responsible Parties (RPs) for oil spills, and to use the recovered damages to "restore, rehabilitate, replace or acquire the equivalent" of the injured resources.

The trustees conduct an assessment to:

- Determine and quantify the injury to natural resources.
- Implement appropriate plans for restoring the resources to prespill conditions.

- Compensate the public and the environment for losses suffered from injury to the natural resources prior to their recovery.[2]

The final rule on NRDA became effective February 1996.

USE vs NON-USE

To better grasp NRDA, a few regulatory definitions are in order. We use the words *damage, injury, restoration* and *natural resources* in daily speech, but these key concepts in the NRDA process differ from their commonly accepted meanings.3

Damages: The amount of money calculated to compensate for injury to, destruction of, and loss of natural resources. In NRDA terms, damages include the loss of use of the resource until its recovery to prespill conditions and the costs of determining the damage to the resource. Only the U.S., a state, tribal nation, or foreign trustee can recover damages from the RP for natural resources.

Injury: Any measurable or observable adverse change in a natural resource. Again this term is comprehensive under NRDA's scheme. Injury includes impairment of a service provided by a resource. Loss can be total, if complete destruction of the resource occurs, or less than irreversible where there is a measurable adverse reduction of a chemical or physical quality or viability of the natural resource.

Restoration: Encompasses actions that accelerate recovery of an injured resource to its baseline condition, including natural recovery; and activities that compensate the public and ecosystem for lost services from the onset of injury until return to prespill condition.

Natural Resources: The definition includes "land, fish, wildlife, biota, air, water, ground water, drinking water supplies, other such resources belonging to, managed by, held in trust by, appertaining to or otherwise controlled by the U.S., any state or local government, tribal nation, or foreign government."

Interim Lost Values: Refer to the reduction in the level of natural resource services, which can be recreational, commercial, cultural or ecological, from the date of injury until the resource is restored to original condition.

Passive Use Values: Part of interim lost values, consisting of values individuals place on natural resources independent of any direct use of that resource by a person. These may include:

Option value, or the value of knowing that the resource is available for use by friends, family, or the general public over and above any planned use;

Existence value, or the value derived from protecting the natural resource for its own sake just because the resource exists; or

Bequest/Heritage value, the value of knowing that future generations will be able to use the resource. A passive value might be what you as an individual are willing to pay to know that polar bears will con tinue to thrive in Alaska, or to ensure that the Grand Canyon will be there for your children and grandchildren to visit, or to guaran tee that coral reefs remain alive and well for the thousands of crea tures who inhabit them well into the distant future.

Douglas K. Hall, assistant secretary for NOAA, provided a wonderful example of the use vs. non-use value of natural resources:

> "Natural resources have both use and non-use values ... Most individuals are willing to pay more to fish a clear mountain stream in the wilderness of Idaho or Montana than to fish in a polluted pond in a denuded wetland adjacent to a chemical manufacturing plant. Non-use values are those not arising from the actual use of the resource. They include knowing that the Idaho stream exists, that individuals will be able to fish that stream in the future, and that the stream will be available for our children and their children."[4]

Native American tribes, one of the trustees recognized by DOI and NOAA, put a different spin on the values issue:

> "It is more accurate to refer to these kinds of losses as 'heritage' losses, or loss of existent value and bequest value ... Our natural resources, and how we use them, define who we are and how we live ... Natural resource injury does not simply mean that 'so-and-so' has to fish someplace else next year. It means that our entire identity, our values, our religion have been injured. This is the significance to the tribes of the existence and bequest value of natural resources. That is why it is essential that these tribal interests in natural resources be recognized and protected in this statute."[5]

Natural resource damage claims under OPA 90 allow for recovery of three basic types of damage:

- The cost of restoring the resource.
- Compensation for interim loss of the resources and/or services pending recovery.
- The reasonable cost of performing the damage assessment.[6]

RECOVERY UNDER THE NOAA REGIME
The NOAA regulations focus on the ultimate goal of OPA 90's NRDA:

" ... to ensure the restoration of natural resources and services that have been injured, destroyed, or lost as a result of an incident." The final rule involves a major shift in emphasis from a valuation-based approach to a restoration-based approach.

Under NOAA's rule, trustees reach their restoration goal through a three-phase process: (1) preassessment; (2) restoration planning; and (3) restoration implementation. In the preassessment phase, trustees determine whether resources have been injured and warrant continuation of the assessment process. This determination is made by finding (1) if the cleanup adequately addressed the natural resource injuries caused by the spill; (2) if additional restoration is necessary to return the resources to their baseline condition; and (3) if feasible restoration alternatives exist. If warranted, trustees proceed to the restoration planning phase.

The second phase involves injury assessment and restoration selection, and links injury and restoration. Injury assessment includes determining the nature and extent of the injuries and providing a technical basis for evaluating what is needed to restore resources. Valuation may only be used to plan the size of restoration actions in order to produce an equivalent to the lost value of the natural resources. To determine injury, the trustees compare the injured resource to baseline conditions, quantify the degree of the injury, and estimate the time until the resource returns to these same conditions.

In quantifying the injury, trustees may use any reasonable assessment method appropriate for determining restoration needs for the incident, so long as the method is "reliable and valid" and cost-effective for the incident. Models like DOI's Type A, compensation formulas, and procedures based on field methods, laboratory methods, model-based methods, and literature-based methods may be considered under NOAA's rule.

THE GOAL: RESTORATION

Once injury assessment is complete, trustees develop a restoration plan. Selection of the restoration alternative is the next area of focus. Trustees must identify a reasonable range of alternatives that would restore the resources or compensate for interim lost services and select a preferred restoration action. *Primary restoration* actions are taken to return the injured resources to their prespill state, which can include natural recovery without human intervention. *Compensatory restoration* actions are taken to make the environment and the public *whole* for the resources and/or services lost from the date of the incident until the date the resources return to prespill conditions.

Using the OPA 90 rule, trustees must scale the restoration actions appropriately. The OPA rule is project-based. Trustees first apply a *service to service approach*, which might include opening a new beach to replace an injured beach while it recovers to baseline, prespill condition. If a *service to service approach* is unavailable, trustees may use the valuation approach to scaling, comparing the value of services lost to the value of services restored. Trustees identify, select and scale actual projects to replace or acquire the equivalent of the lost uses. The cost of implementing the project is part of the damage claim. Contingent valuation and other valuation methods become important only to the extent that they identify the appropriate scale of natural resource restoration projects.

No longer is the dollar value the end point of the assessment approach. NOAA's law sets up a preference for scaling compensatory restoration projects according to the amount of services they provide, rather than solely measuring the lost/returned human dollar values.

Trustees must select a preferred alternative from a reasonable range of feasible restoration actions, and consider solutions based on these important factors:

- Cost of the plan.
- How well the plan carries out the goal of recovery.
- Likelihood of success.
- If the plan avoids additional injury.
- Benefits provided to more than one resource.
- Effects on public health and safety.

If two alternatives are equally preferred under this analysis, the trustees are charged with selecting the most cost-effective option. The draft restoration plan is then made available for comment to the public and scientific community.

In the third and last phase of the NOAA assessment process, trustees present the final restoration plan to the RP for comment, allowing the RP an opportunity to settle claims without litigation. If the RP rejects the plan, the trustees can pursue damages.[7]

MODEL RECOVERY

DOI designed two computer-generated models for what they refer to as *Type A* assessments, one for the Great Lakes environments, and one for coastal and marine environments, in response to OPA 90's intent to minimize costs and efforts expended in the assessment process, particularly in the case of small spills. According to studies of coastal discharges between 1973 and 1990, 99.8 percent of all oil discharges were of less than 50,000 gallons, and 99 percent involved spills of less than 10,000 gallons. These are small spills compared to the 798,000 gallons released in *Berman*.[8]

The authors developed a set of candidate spill occurrences based on historical records, using spill data provided by NOAA which was originally obtained from the USCG's Marine Pollution Retrieval System. They used 337 spill event records, considered to be representative of coastal spills annually. Essentially, the computer models produce a dollar value for each gallon spilled, and damage estimates depending on the region of the spill. The NOAA final rule allows use of these Type A models so long as they meet certain criteria, including reliability and validity.[9]

CONTINGENT VALUATION/WILLINGNESS TO PAY

NOAA considers Contingent Valuation (CV) as one method to fix the monetary loss sustained by the damaged natural resources. This is one of the most controversial pieces of NRDA, regardless of whether the DOI or NOAA rule applies. The international community and the U.S. butt heads in this area.

The Contingent Valuation Method (CVM) is a survey-based approach to determine what the public is willing to pay to preserve the existence of a natural resource, or is willing to accept as compensation if the resource is damaged.

> "Contingent Valuation is a survey technique using direct questioning of individuals while they are on-site or by mail to generate estimates of individuals' willingness to pay for something they value ... Alternatively, individuals might be asked how much compensation they would require if they no longer had access to the natural resource."[10]

In CV, an assessor establishes the pool of people affected by injury to the resource, tries to select a representative sample of people from that pool, determines through an interview process the average value they are willing to pay to prevent damage, then multiplies that average by the total population in the affected pool. The end product is a specified amount of damage. In the case of *Berman*, almost any person in the U.S. might be a member of the pool, since people across the country value and travel to the white sand beaches and clear Caribbean waters of this U.S. commonwealth.

NOAA sought the advice and counsel of a panel of experts including Nobel laureates to develop a regulation concerning the use of CV and proper methods for assuring the greatest scientific reliability in valuing non-use natural resource lost services. This panel found that CV is a reliable method for determining lost passive use value as long as the trustees follow certain guidelines and use a conservative approach. The method designed must *understate* rather than overstate the resource damage.

The panel was convened for many reasons, among them legal concerns. In a court of law, if the statement that "a coral reef is priceless" is the only *evidence* presented, a judge is quite capable of awarding the trustees and public a zero judgment. Numbers must be placed on damaged resources for litigation and settlement contexts, and the system demands some reasonable method to establish lost value. The panel devised strict guidelines to increase reliability of the CV method.

The panel recommended common-sense limitations to increase the reliability of CV surveys, and concluded that CV studies can produce good estimates which include lost use values that serve as a starting point in a judicial damage assessment. The NOAA panel did not require complete adherence to every guideline for a CV determination to be considered reliable, but cautioned against too many departures or deviations from the numerous guidelines in its published report. In summary:

> "These require that respondents be carefully informed about the particular environmental damage to be valued, and about the full extent of substitutes and undamaged alternatives available. In willingness to pay scenarios, the payment vehicle must be presented fully and clearly, with the relevant budget constraint emphasized. The payment scenario should be convincingly described, preferably in a referendum context, because most respondents will have had experience with referendum ballots with less-than-perfect background information. Where choices in formulating the CV instrument can be made, we urge they lean in the conservative direction, as a partial or total offset to the likely tendency to exaggerate willingness to pay."[11]

The panel's conclusions about CV reliability were endorsed in a 1996 study by Resources for the Future, a Washington, D.C.-based think tank.

RESTORATION: A SUCCESS STORY?

After undertaking the careful analysis described above, while not specifically authorized under NOAA's rules, trustees may elect to pool recovered funds and undertake a Regional Restoration Plan on a geographical or habitat basis, where that alternative represents a more environmentally effective use of damages recovered. The plan must address similar or same resource injuries as those identified in the assessment.

Where might this be appropriate? Assume an oil spill kills seagrass beds in an affected area, such as what happened in *Berman*, where no amount of care could save the seagrass blades in an extensive area oiled by No. 6 diesel fuel. Blade die-off was substantial. Time is needed for new blades to grow. Under the international regimes, no replanting of seagrass beds offsite would be allowed to compensate for lost productivity of the injured beds while they recovered from blade die-off. NOAA takes a different approach. If a seagrass bed in an adjacent bay along the same coastline has been polluted for years by waste discharge, funds recovered from the RP can be used to clean up and prevent further sewage contamination to this nearby site so the restored seagrass bed can become a healthy home for marine organisms displaced by the spill.[12]

According to NOAA, their restoration actions are successful. "Working with co-trustees," states the agency, "NOAA has recovered more than $140 million in restoration funds with an investment of just under $10 million in appropriated monies. This represents a return of more than $10 for the restoration of injured coastal and marine habitats for every tax dollar spent."[13]

ITOPF holds the short-range view that replacing a damaged habitat with an equivalent resource is ridiculous. If the resource in question is gone, it is gone, and the polluter has no obligation to clean up someone else's mess.[14]

REBUTTABLE PRESUMPTION

Under OPA 90, if the trustees determine natural resource damages in accordance with the final rule promulgated by NOAA, the resulting damage assessments are entitled to a *Rebuttable Presumption* valid in any administrative hearing or judicial proceeding under OPA 90. Some who work in the system believe this gives the trustees an advantage. Others regard this presumption as an almost impossible impediment for RPs. NOAA interprets this presumption to mean that "the RP has the burden of presenting alternative evidence on damages and of persuading the fact-finder that the damages presented by the trustees are not an appropriate measure" of the natural resource injuries caused by a spill. Still others believe that rebuttable presumption is meaningless.[15]

NRDA UNDER ATTACK

NOAA's January 1996 final rule is the subject of six petitions or lawsuits filed by General Electric Company (GE suit), American Institute of Marine Underwriters and Water Quality Insurance Syndicate, the Chemical Manufacturers Association, the American Petroleum Institute, Beazer East, Inc., and Zeneca Holdings, Inc. Five others have asked to join in this suit, including shipping industry giants such as the International Group of P & I Clubs, who claim to insure over 95 percent of the world's ocean-going merchant fleet, including effectively all oil tankers, and Underwriters at Lloyd's of London, who state that their underwriters collectively write 15 percent of the worldwide marine insurance premiums.

These companies seek to invalidate the guts of NOAA's final regulation, and specifically object to:

- Trustees' use of uncontrolled discretion in assessing natural resource damages and selecting the restoration alternative.

- Failing to make cost-effectiveness the overriding concern in choosing a restoration alternative.

- Allowing damages for lost use without requiring strict accounting for the service provided by the natural resource.

- Allowing use of CV as a technique for measuring non-use values of injured resources.

- Including non-use values as a basis for scaling restoration.

- Authorizing trustees to use the Type A DOI model.

- Other problem areas, such as the existence of a rebuttable presumption in NOAA's favor, allowing regional restoration plans, and using estimates to value lost or impaired services.

At the other end of the spectrum, the Natural Resources Defense Council is one of the five parties asking for leave to join this suit, asserting that the final rule does not go *far enough* in environmental protection and restoration.[16]

Terry Garcia of NOAA's general counsel answers these challenges with confidence:

"NOAA will stand behind and defend fully the final natural resource damage assessment rule under the Oil Pollution Act of 1990. We will defend this rule under the Administrative Procedure Act. It is consistent with the intent of Congress as demonstrated in OPA."[17]

The NOAA final rule received a major boost when the same court upheld the DOI March 1994 NRDA regulation against industry attack in the case of *Kennecott Utah Copper Corporation, et al v. U.S. Dept. of the Interior*, No. 93-1700 (and consolidated cases) U.S. Crt. Of App., (D.C. Cir.) (Kennecott suit) in a decision announced July 16, 1996. Even according to the petitioners in the GE suit, many of the issues raised in the GE suit are "substantially similar" to those attacked in the DOI suit and will "either resolve or significantly affect" the outcome of the GE suit.[18]

TWILIGHT ZONE

NRDA underscores a deep conflict between science and law that has caused some parts of the scientific community to throw up their hands and condemn the U.S. approach toward environmental damages.

In the area of evidence, scientists function in what has been called a *twilight zone* between experimental evidence and demonstrable proof. The test for good science is *a degree of statistical certainty*: "The scientist will strive for the statistical evaluation of a given set of data to prove to a specified degree of mathematical probability that an event was not due to chance or an alternative hypothesis."[19]

In NRDA, this *scientific* measure yields to the *legal* definition of proof, under the civil standard of what is "more likely than not." "In other words, the trier of fact may find that an event occurred because they believe it is more likely to have occurred than not to have happened at all."[20] Scientists call the legal decision a "gut level" determination.

There are those involved with NRDA who see the goal of U.S. environmental laws as restoring the environment, and urge the community to stop debating between science and law and to focus instead on restoring and repairing the damage man and his oil needs have caused to natural resources.

Tom Campbell, a lawyer with Entrix, a major industry evaluation firm for oil spills, exemplifies this view:

> *"What's needed in the restoration area is flexibility and creativity. We can't arrogantly assume that science can solve every problem man creates in nature. Sometimes the best we can do is enhance nature's ability to recover. Scientists can't continue to look myopically at a specific insult. They need to look ecologically at the whole system, define what's wrong, and try to fashion a solution that helps. If we are overly arrogant and insist on correcting everything, we'll never reach an acceptable plan. We need to assess the injury and then act, move on to the restoration phase."*[21]

SHIPPING TO THE U.S.: AN EXCLUDED RISK

The U.S. potential for unlimited liability and the range of damages for natural resources strikes at the heart of the shipping industry's Protection and Indemnity (P & I) Club philosophy. The P & I Club is a non-profit mutual association of shipowners who collect sufficient monies to meet calls for collective liability in a given year. P & I Clubs have reached a critical point concerning the U.S. regulatory policies, teetering on the line between maintaining coverage for some shipowners and what they see as a violation of their obligation to all other members.

ITOPF's Nichols summed up the primary issue facing the insurance industry as: "Whether exposure to environmental damage resulting from the carriage of oil to the United States has become so extreme that it, too, should become an excluded risk."[22]

NRDA provisions trouble the industry so deeply that P & I Clubs threaten to refuse insurance coverage of oil trade to the U.S. Starting with *assessment*, NRDA's foundation, the general industry opinion is that *all* NRDA methods are highly specious.

> *"Given that the natural resources involved in environmental valuation exercises are not commercially exploited and do not therefore have a market value, it follows that all methods which have been developed are highly theoretical and speculative, and give inconsistent and arbitrary results."*[23]

Industry claims that specific DOI models fall short of the scientific certainty and reasonableness industry standards demand. Bessemer Clark, an expert retained by ITOPF, states a startling example:

> *"The costs revealed by these formulas are real enough; in many cases they are likely to be huge: An offshore winter spill of a single barrel of heavy crude in Southern California yields an NRDA assessment of $279,686; in San Francisco Bay in summer it would be $1,162,967."*[24]

No supporting data has been supplied for these amazing figures, despite request. In response, DOI's James Bennett, an environmental specialist who had a primary role along with others from DOI in developing the DOI models, believes that these figures were generated from earlier NOAA formulas no longer in use, and that if the figures were plugged into DOI's final model, they would not generate such a sum."[25]

Dr. I.C. White, managing director of ITOPF, asserts the international view that (1) the marine environment can recover, often without the help of man; (2) oil discharges have short-term, impermanent effect on the environment; and (3) prespill conditions cannot be achieved. There are limits to what man can repair, and natural recovery is preferred.[26]

Richard Hobbie of the Water Quality Insurance Syndicate warned the U.S. at hearings before Congress in 1995 that natural resource damages might become "uninsurable," causing his company to cancel all financial guarantees issued to vessel owners under OPA 90.[27]

NRDA proponent Sarah Chassis of the Natural Resources Defense Council answers that if NRDA provisions of OPA 90 are eliminated, "either U.S. taxpayers would have to shoulder this expense, or the damage to public resources would go unabated, leaving a toxic legacy for future generations."[28]

THE U.S. AT WAR

The battle has shifted from the international front to the halls of Congress, where members have attacked the basic concept that the polluter pays. In 1995, legislation was proposed in both the U.S. House and Senate intended to eviscerate the Comprehensive Environmental Response, Compensation and Liability Act of 1990, known as the Superfund Act. OPA 90 NRDA provisions are one victim of the proposed House bill. Highlights of the attack on OPA 90 in the House are:

- Defining all restoration plans in terms of cost-effective and timely action.
- Eliminating recovery for non-use damages.
- Taking away trustees' rebuttable presumption.

While many believe these industry attacks will fail, powerful oil, chemical and other commercial interests have clearly mobilized an effort to revise OPA 90. The future will see repeated attempts to limit recovery for damages to natural resources, and to undermine the current *polluter pays* philosophy of OPA 90 and CERCLA. To paraphrase the Senate testimony of one NOAA official, we must not allow the interests of the American people to be subordinated by the interests of the polluter.[29]

TRAIN WRECK AVERTED

The international shipping community's attitude gives cause for alarm because the U.S. depends almost exclusively on oil imports through ship transportation, and ITOPF (through its P & I Clubs) insures a high percentage of the shipping industry. The U.S. faces the real future threat of an international shipping boycott.

In one aspect, a *threatened* boycott already happened. The international insurers dug in their heels and said "no" to the financial risk policy implemented under OPA 90 Certificates of Financial Responsibility (COFR), required for every shipowner trading in U.S. waters. The P & I Clubs threatened that international ships would not transport oil or commodities in U.S. waters. Says Nichols:

> *"The real issue is not so much OPA 90 as it is the new Certificate of Financial Responsibility the U.S. is now requiring. Trading ships have been carrying certificates under the CLC, issued by the Flag State, to show that the vessel has sufficient insurance to meet its oil pollution cleanup requirements ... The coverage was about $83.6 million.*

> *"In September 1994, the USCG issued a new regulation pursuant to OPA 90 changing the whole scheme. Ships are now required to have insurance in place up to $1200 per ton, the OPA limit. That's okay. We can live with increased 'limits.' The problem is this stated amount isn't a limit. Under the new plan, because of the potential for unlimited liability, the insurance company ends up being an indemnity guarantor. That means that the shipowner pays for a ridiculous range of damages, the club indemnifies the shipowner, then the claimant or person harmed can go directly against the club.*

> *"This has never happened before, and puts a P & I Club in an unacceptable position of being the victim of class actions for example by every fisherman anywhere near the spill vicinity and suits for unlimited amounts of damage. The P & I Clubs simply won't continue to insure ships trading to the U.S. Tonnage rates are so low, shipowners may well decide not to traffic in the U.S."*[30]

This is the dire *Train Wreck scenario* predicted by the international insurance industry.

Since 1970, the USCG has required vessels operating in U.S. waters to evidence financial ability to comply with rules regulating discharges into navigable waters under the Federal Water Pollution Control Act (FWPCA). Under OPA 90, a new COFR will eventually replace outdated FWPCA certificates and also meet previous CERCLA requirements.

The USCG has set a phase-in period to help current vessels meet the new increased insurance and other requirements: tank barges, non-self-propelled vessels, and all tankers must have COFRs in place currently. All other vessels can use their old certificates until they are due for renewal or until December 28, 1997, whichever comes first. This longer phase-in period applies mostly to cruise ships and other vessels not carrying oil, to force the largest risk vessels to comply first.

The COFR increases the limits of potential liability for any vessel. For a tank vessel over 3000 gross tons, limits are the greater of $1,200 a gross ton or $10 million; and the greater of $1,200 per gross ton or $2 million for those ships of 3,000 gross tons or less. For any other vessel the shipowner must have insurance coverage of $600 a gross ton or $500,000, whichever is greater. A 100,000 gross ton crude oil tanker might incur liability of up to $150 million under the new limits.[31]

However hefty the increased limits may seem, if these amounts were all that was required, the insurance industry would live with them. The main problem from the insurers' point of view is that the same provisions in OPA 90 that allow liability to be breached also apply to the new COFRs. For any of the following conditions, the shipowner may be subjected to *unlimited* liability:

- Gross negligence.
- Willful misconduct.
- Violation of any federal safety or regulatory law.
- Failure to cooperate in the cleanup of any actual or threatened discharge.
- Failure to report the incident.

The new COFRs extend the right of suit in an oil incident to anyone, not only the U.S. government, as before. This includes every state that has its own pollution regimes, and any individual harmed by the discharge. In practical terms, if a spill occurs, the operator pays the cost, the P & I Club reimburses the operator, the vessel owner is indemnified by the P & I Club, and the P & I Club is left hanging for possible unlimited amounts of damages with a right of direct action by any individual or entity sustaining damage against the Club. P & I Clubs can no longer assert insurance policy defenses to limit their coverage or liability.[32]

This is stringent legislation. But the U.S. has a reason for placing the American consumer in danger of a boycott by oil shippers:

> *"What we're trying to accomplish is not allowing any vessel into U.S. waters without the ability to pay in case that vessel causes a spill. Anyone who doesn't have a certificate showing the shipowner can pay up to the statutory limit for any expense involved in dealing with an incident and for damages to the full extent of the law won't get in. If P & I Clubs won't issue the new insurance, a shipowner can meet the requirements by self-insuring, putting up a surety bond, or getting someone else to insure his vessels."*[33]

The assistant chief of the National Pollution Fund Center's (NPFC) Vessel Certification Division summed up the U.S. position:

> *"Nationwide there's a system that tracks all certificates issued to all vessels. That's the Marine Safety Information System (MSIS) USCG computer network. Vessels of a given size have to give 24-hour advance notice to the local Marine Safety Office before arriving in port. The MSO can check on the computer network to see if the COFR has issued or is in the process of being issued and the paperwork is all that's pending. If there's no certificate, the ship simply can't come into the port. The international community might not like this scheme, but it's here to stay. Congress isn't changing the law."*[34]

One can only hope that Armstrong's statement will hold true in the present and future political climate.

To put the COFR situation into perspective, consider that the Berman/Frank family walked away from the massive *Berman* spill. Their only obligation was $10 million in insurance. Until huge fines from criminal actions filed against certain of their companies are paid, the U.S. consumer picks up the tab. Under the new COFR requirement, the companies involved would either have had sufficient insurance to meet the cost of the $87 million cleanup and an untold amount of subsequent damage, or their ships would not have been allowed into the San Juan Harbor. Given the Berman/Frank family's penchant for shaving costs to maximize profits, the *Berman* spill might not have happened if the COFR requirement applied to tugs and barges and was in effect in January 1994.

THE ANSWER: SUPPLY AND DEMAND

Has the international community refused to transport oil to the U.S. in a move to boycott COFRs? Of course not. Shipowners, like the rest of the commercial world, follow the rule of supply and demand. Where there is a commercial need, someone will supply the product or service.

There was no *train wreck*. As of March 7, 1996, the USCG issued COFRs to more than 5700 tankers and tank barges.[35]

Before OPA 90, there were substantial gaps in the liability system. A major oil company with huge profits could afford to pay the spill damages due by law and absorb the loss as a cost of doing business. OPA 90 sought to plug the gaping hole between oil transportation and its detrimental impact on the environment. The question is whether OPA 90 has gone too far, or not far enough. The new U.S. legislation of higher limits, loopholes in liability, insistence on compensation for natural resources, and COFRs is still in its infancy. The benefits to people and to the environment remain unknown.

As oil consumers and ultimate payor, we have the right to demand that oil travels first class. Those who do not want to transport petroleum with the respect such a highly dangerous substance deserves should avoid shipping in our waters. If the rest of the world would join our regimes rather than oppose them, the world's oceans could be made much safer. What happens elsewhere at sea ultimately impacts our own waters. The U.S. has a responsibility as the largest oil consumer to demand compliance with the most innovative and environmentally sensitive legislation on the books. All we require is the political will to stand behind our own laws.

PREVENTION THROUGH PEOPLE

Berman raises serious questions about the shipping business. With all the regulations, laws and controls nationally and internationally, how could such an incident happen? More importantly, what can we do to *prevent* future *Morris J. Berman* spills?

Berman represents a startling truth. Despite the numerous safety violations involved in *Berman*, *human error* is the real cause in 60 percent to 80 percent of all oil discharges.[1]

In the past, the shipping industry has concentrated on technology to promote safety and minimize marine casualties. *Berman* and other large oil spills of the 1990s make it painfully clear that engineering and technical solutions no longer work. Men, not machines, are the real line of defense against maritime accidents. Internationally and nationally, the maritime community is shifting its dependence from steel to people.

This new focus on people is creating measurable change in six major areas in the marine environment:

- Reform in the USCG licensing procedures for all crew on board tankers and other vessels, as envisioned by the USCG's seminal document *Licensing 2000 and Beyond*, with reforms paralleling the STCW International Convention changes.

- IMO's new mandatory International Safety Management Code (the ISM Code) designed to bring about safer management policies and practices in companies transporting oil across the world's seas.

- Industry changes patterned after the ISM Code, such as the Oil Companies International Marine Forum (OCIMF) database, certification programs and internal auditing procedures established by Texaco and other large conglomerates.

- Institution of a USCG point system for vessels entering U.S. ports under the right of a Port State to control its own waters.

- The global movement by other countries to regulate the safety of foreign vessels entering their waters through regional and international Port State control programs.

- Unique conventions such as Intervention on the High Seas, which authorizes the ultimate control over a foreign vessel endangering a country's waters.

THERMOS BOTTLE

Within this important shift in emphasis in the shipping industry from engineering to people-driven solutions, one technological change will have a particularly far-reaching effect on the safety of tankers carrying oil. OPA 90 mandates that *by the year 2015* all oil carried in U.S. waters must be shipped in tank vessels with *double hulls*.

The barge in *Berman* was a single hull. When the barge struck the reef off Punta Escambrón, only one protective skin separated oil from ocean. The reef pierced this skin, and the tanks emptied into the sea. A double hull works much like a thermos bottle, insulating the cargo in two layers. The difference is dramatic. According to one source, in *Berman* the tanks with minor damage would not have spilled oil into the sea if the barge had been of a double hull design.[2]

The anticipated goal of the double hull requirement is to reach *zero probability of oil outflow* by the best technical means man can devise. Not until *Exxon Valdez* did Congress feel impelled to take the plunge after 15 years or so of debate about double hulls, and ordered the shipping industry to replace its fleet:

> *"Up until now, the additional cost of double hulls has been the deciding factor against their implementation. But compared with the monumental cost of cleaning up Prince William Sound, as well as the potential for devastation posed by groundings such as the recent one off the Florida Keys, the increase in construction cost is minor.*
>
> *"We must tackle environmental questions before they threaten the public health and safety, not afterward.* We have learned in Alaska that you can't put the genie back in the bottle (emphasis added). *When the bottle, in this case the Exxon Valdez, weighs 126,000 tons, it is easy to see that the cost of prevention outweighs the priceless loss in environmental damage from another invasive spill.*
>
> *"I do not want some future headline to read 'Washington Studies While the Oil Spills.'"*[3]

Even in the face of an 11-million-gallon oil spill, U.S. politicians gave into lobbying pressure and compromised. Section 4115(b) of OPA 90 provides for a phaseout period for replacement of current vessels. Any vessel constructed after June 30, 1995 must be double-hulled. All other vessels must be replaced or rebuilt depending on their age and size (tonnage), with the oldest and least environmentally sound vessels being phased out beginning in 1995.[4] OPA 90 charged the USCG with instituting structural and operational measures to upgrade single-hulled tankers until implementation of the double hull measures.

Finally, on August 5, 1996, the USCG published its final rule,

Operational Measures to Reduce Oil Spills from Existing Tank Vessels Without Double Hulls. This rule addresses certain operational requirements for non-double-hull tank vessels, including barges of 5,000 gross tons or more – sorely needed, as *Berman* shows. Highlights include:

- Measures to reduce human error by promoting communication between the tank ship's master and local pilots and between masters and officers in charge of the watch.
- Enhanced maintenance surveys and maneuvering capability testing.
- A requirement that the tank barge owner or operator ensures that the towing vessel operation meets the same new and increased standards required for tank ships.
- Imposing under-keel clearance guidelines.[5]

Double hulls may come into existence in U.S. waters sooner than the year 2015 if the most recent legislation passes. Sen. Chafee reacted to the disastrous oil spill of 828,000 gallons of No. 1 fuel oil off the coast of Rhode Island in January 1996. While his bill may not pass in the 1996 session, this bill highlights the direction the U.S. may take: shippers who elect to double hull their fleet by the year 2010 will be rewarded. Their liability limits for oil spill damage will not be subject to current liability exceptions unless there is gross negligence or willful misconduct involved in the spill.[6]

The international community has not adopted the double hull standard wholeheartedly. Again, the U.S. as a leader in the maritime field imposes a higher design standard than other countries appear willing to accept. After OPA 90, IMO adopted an amendment to the MARPOL Convention calling for a double hull or "equivalent pollution prevention" design for tankers older than 25 years to cover at least 30 percent of the cargo tank area.

An additional IMO change requires tankers built in 1987 or later to undergo "enhanced inspections" to check for structural deterioration. SOLAS amendments mandate elevated inspections for bulk carriers, effective January 1, 1996. IMO believes these enhanced surveys will subject more than 80 percent of existing tankers to greater scrutiny than ever before. The intended effect is that owners will choose to replace aging tankers with new tankers as increased inspections, required conversion costs and repairs make running older ships less profitable.[7]

Admiral William Kime, former Commandant of the USCG, addressed IMO and spoke about the heated debate between the U.S. and the rest of the world, to which double hulls contribute further ill will:

"Existing IMO requirements are not properly enforced by all Flag States. Changes will be necessary. The old way of doing business will not be acceptable any longer. These changes will cost money ... However, if the IMO truly responds to these issues in a practical, effective, cost-effective and timely manner – not with patches, quick fixes and a least-common-denominator compromise of business as usual – then the clouds of unilateral action by both individual countries and regional groups will not rain on the maritime industry. While OPA 90 was born in Washington, it was conceived here in London due to slow and inadequate action."[8]

HUMAN ERROR/HUMAN FACTORS

Design alone will not prevent accidents. *Exxon Valdez* was a state-of-the-art tanker, yet human error brought her onto the rocks. *Braer* was a ship involved in a high-energy environment disaster, where no hull, single or double, could withstand the repeated pounding of steel against rock. People enter the equation. The most advanced technology fails when humans err.

When Rear Admiral James C. Card assumed his position as the USCG chief, Office of Marine Safety, Security and Environmental Protection, he announced a new era in U.S. seafaring policy:

"Despite (these) engineering and technological innovations, significant marine casualties continue to occur ... There is a clear need to critically address people issues. The issues must be addressed, not only from the traditional man and machine interface and economics aspects, but must also include an assessment of entire processes including navigating the vessel, cargo loading and unloading, and responding to emergencies."[9]

Why this philosophical shift from steel to man? Studies conducted by the USCG in the wake of the Amtrak *Sunset Limited* incident near Mobile, AL and the *Berman* spill revealed some startling information: man, not his machines, were the cause of the majority of oil discharges. A USCG study into the source of reportable marine casualties between 1982 and 1991 resulting from operation of uninspected towing vessels (UTV) like the tugboat *Emily S* involved in the *Berman* spill concluded:

"The results show that of the reported UTV casualties, approximately 62 percent were caused by human factors, approximately 18 percent were caused by equipment and material failures, and approximately 20 percent were caused by environmental and other factors. Three major causes of human factor casualties were found to be: oper-

ator error, error in judgment, and improper action. Carelessness, maintenance, training and other activities (including alcohol and drug use) accounted for seven percent of the human factor causes."[10]

Other studies find that human errors account for up to 80 percent of all marine casualties. "These analyses indicate that 65 percent to 80 percent of casualties are caused by people."[11]

The USCG has defined *human error* as acts or omissions of personnel which affect successful performance. *Human factors* involve the interaction between equipment and the human operator, and most importantly, the procedures the crew and management follow.[12]

The USCG maritime community is shifting its focus to human error. "The purpose of studying human factors is to identify how the crew, the owner, operator, the classification societies and the regulatory bodies can each work to sever the chain of errors which are associated with every marine casualty."[13] Two major areas are targeted: licensing of crew and their training, including competency requirements and testing for initial certification or recertification of crew members; and control over vessels entering U.S. ports. Man and his technological systems do not operate independently of each other. Decisions made at the highest ranks of a company affect those at the bottom of the chain – the crew who run the ship. *Berman* is a case in point. The Berman/Frank philosophy of shopping for the best buy, of waiting for that one last trip before changing out a frayed towing wire, translated downward through their local agent, the captain of their tugboat and the mate at the wheel during the last hours before the wire finally parted.

At almost every level in *Berman*, better management, maintenance and training in emergency procedure, insistence upon sleep, more careful attention to the vessel, and less concern for profits could have avoided the disaster. Instead, the Berman/Frank companies ignored these people-oriented controls, and oil spilled into the ocean.

LICENSING PROCEDURES

How can we prevent humans from making the same mistakes that occurred in *Berman*, and that apparently happen in 60 percent to 80 percent of all major spills? One way is to make an example of those in charge after the fact. This means stiff civil penalties against the shipowners and their top management, criminal actions against owners for failure to comply with safety rules, casualty investigations and revocation of licenses for incompetent or grossly negligent masters and their mates, and unlimited liability actions that hold each Responsible Party strictly negligent for every penny of cost.

In addition to the penalties imposed on *Berman* participants described in Chapter 18, assets of spillers may be seized and sold. The commonwealth of Puerto Rico sold the *Berman* tug *Emily S* for $750,000.[14]

The serious legal actions taken after *Berman* send a strong message to shippers that they must use care in transporting oil because they will be held personally liable for violations. But these penalties are after the fact. The revised U.S. policy concentrates on before-the-fact prevention and begins with the people who command the ships and those who respond to their orders. The intent is to create a competent and careful work force – an especially difficult task during a time of budget constraints, cost-driven considerations, and smaller, more overworked crews speaking multiple languages.[15]

Licensing of mariners is the key. Our current licensing system is way off track. How we became so derailed can be understood by a glimpse into the history of the shipping business.

In the old days of wooden sailing ships, when the captain maintained sole rule over his vessel, there were no licensing requirements. According to Capt. Ross, the sea took care of the job by a process of natural selection. If you survived, you learned how to maneuver your vessel in the safest and most effective manner. Society did not bear the brunt of your failure. The worst that resulted from ineptitude was the loss of your cargo, what Ross refers to as some "bad French wine."

As shipping became more important to the social fabric, so did the need for protection of merchandise and the consumers who ultimately enjoyed the results. Insurers like Lloyd's of London asserted a form of regulation which included licensing of ship's masters. Given the rigors of their environment, the apprentices who served under these masters of the sea either learned or died. As the regulatory process evolved, these same master mariners administered oral examinations for licensure. Their careful subjective review, along with demonstrable sea experience, guaranteed a high level of competency among crews navigating the oceans.

This old system of verifiable and meaningful tests and sea experience gave way to budget-driven considerations. Multiple-choice, question-and-answer examinations administered by file clerks replaced oral tests given by master sailors. Training courses sprang up, intent on helping those seeking licensure to cram for exams. First-hand sea experience judged by highly skilled seamen fell by the wayside. Quantity of time replaced *quality* of time and training spent on the open water.[16]

The end result is a licensing system that is broken and beyond repair. Tug captain McMichael and his mate were on their fourth period of relicensing by the USCG. The captain did not even know the basic rudiments of emergency cable splicing, and his first mate was too tired or too drugged to keep a proper watch. *Berman* happened.

ROAD MAP FOR THE FUTURE

Licensing 2000 and Beyond is the USCG road map for the future. The recommendations in this report will have far-reaching effects on the way U.S. mariners acquire their training and prove their competence now and into the next century. The 46-page report contains an in-depth analysis of what is wrong with the present system, and has a general comment section followed by a specific discussion of what is needed for each category of sea personnel. To those in the business of licensing, this small document has become the manual for success.

Competent mariners is the goal of *Licensing 2000 and Beyond*. *Competence* in USCG lingo means possessing the "total set of skills, knowledge and judgments necessary for the proper performance of one's duties in a specific position on a specific vessel."

Licensing 2000 and Beyond confirms Capt. Ross's opinion that the current licensing system does not work. The report found that competency cannot be shown simply by a set number of "sea years" and successful completion of a written exam. New and improved methods are necessary to determine if a candidate should be licensed or reissued a new license. Training by competent mariners in courses with pre-approved USCG curriculum in a wide range of subjects including modern technological equipment and simulated experience is as important as, and in some cases may replace, actual sea experience. The report emphasizes the *quality* of the training and subjective review of the applicant rather than performance on a multiple-choice exam.[17]

The Focus Group that produced the report recommends adopting a program like the international Model Course Program proposed by IMO under the Standards of Training, Certification and Watchkeeping for Seafarers Convention (STCW), 1978. The IMO course is based on the premise that increased professionalism will decrease the potential for casualties. The model emphasizes training programs, approved courses, designated and licensed examiners and tests which measure practical application. Under the USCG plan, training for American mariners will be handled by the private sector, with the USCG supervising course curriculum and its administration by private schools. The report recommends a mix of comprehensive training and quality sea service.

MORE THAN "BELLS AND WHISTLES"

The views of those charged with making the changes recommended by *Licensing 2000 and Beyond* provide insight into its value. The mission given CDR Scott Glover, former assistant chief of the Merchant Vessel Personnel Division, USCG in Washington, D.C., was to carry out the objectives of this report and ensure that training received in pre-approved USCG courses is meaningful and actual. Says Glover:

> *"We are following the lead of the international maritime com-*
> *munity in focusing on training that is real and verifiable by good*
> *oversight on our part. The IMO has recommended amendments to*
> *the STCW which will set standards of qualifications for mariners,*
> *licensed and unlicensed, and specify the subject areas for training*
> *and qualification without specifying the details.* Licensing 2000
> and Beyond *is our Bible in this area."*[18]

Private industry joined in the process. There are now 300 to 400 approved USCG courses offered by 100 different schools for license applicants in order to satisfy their training requirements and, in certain cases, part of their sea experience.

Substituting some actual experience with classroom courses works as a return to the old tried and true method of practical sea experience supervised by a sharp-eyed captain. Qualified personnel teach pre-approved, highly specific syllabi, personally observe and judge the qualifications of the applicant for license, and administer comprehensive tests that require real knowledge and not just rote memory. Courses range from First Aid, CPR, radar and diesel mechanics to simulator training.

Indications are that modern wonders of technology such as simulator training do work. Glover says:

> *"The Natural Resource Council did an overall study on the use of*
> *simulators for training, exam or replacement of sea service for mariner*
> *applicants. The study found that if properly applied and used, simula-*
> *tors can impose on the student situations they wouldn't have the*
> *chance to experience in 10 years at sea. These machines aren't just*
> *'Bells and Whistles.' They are primary navigation and steering aids,*
> *for example. The experiences run the gamut. There are full bridge sim-*
> *ulators for masters of the ship and simulated hookups for tanker men.*
> *A simulator is far removed from your average video game."*[19]

Glover's office oversees the trainers. A select cadre of personnel make unannounced visits to various industry courses to determine if the instruction is in compliance with USCG regulations. Glover found that 70 percent of the companies are meeting USCG requirement. Those not following the syllabus, failing to give tests, or otherwise found cutting corners receive a written warning and a deadline to correct deficiencies. Failing that, their course is shut down.

The issue of licensing requirements and emphasis on training is still open to debate. Glover adds:

> *"I believe in the new focus on training and improved methods of*

testing. This shift in the USCG focus will affect everyone from the 'Rolex Operators' who own pleasure boats and don't know the first thing about how to operate them to the Master Unlimited License. Each of these individuals will receive the training they need to meet minimum and above minimum competency, and they'll receive it from qualified individuals who can best judge that they learned what they need to know for safe operations."[20]

Others take a slightly more jaundiced view. Notes Capt. Ross, "Watching a new bureaucracy function is like watching elephants mating. It's a noisy, violent, high-profile process. Then you have to wait 18 months to see the results."[21]

INTERNATIONAL TRAINING TREATIES

In a parallel process, the international community is casting its own searchlight on the *people* part of the marine equation. The Secretary-General of IMO made an impassioned plea in 1995 for a shift in attitude among those involved in the shipping industry in his World Maritime Day address:

"We have become so used to the risks involved in seafaring that we have come to see them as a cost that has to be paid, a price which is exacted for challenging the wrath of the oceans. We must change this attitude, this passive acceptance of the inevitability of disaster. When a ship sinks, we should all feel a sense of loss and failure, because accidents are not *inevitable. They can and should be prevented.[22]*

The international maritime community takes this charge seriously. Three major treaties control training and manning in the global transportation of oil: STCW, SOLAS and ultimately MARPOL. Each of these is the subject of current amendments.

In the training area, power over certification and teaching of mariners rests primarily in the International Convention on Standards of Training, Certification and Watchkeeping for Seafarers (STCW), 1978. One hundred and eight countries are signatories to STCW, making the convention effective for almost 93 percent of the merchant shipping tonnage. Significant revisions to the 16-year-old Convention will take effect by February 1997.

STCW will have its own Seafarers' Training, Certifications and Watchkeeping Code, the technical guts of the convention. Part A of the Code contains mandatory standards and Part B contains recommended guidelines. The Amendments set standards for qualifications of all levels of mariners and establish the general subject areas for course study, leaving the details to be designed by the countries and nations (States) who are parties to the Convention. Like the U.S. program, the international focus emphasizes actual demonstration of skills and schooling. A Model Course Program envisions applicants completing approved courses, including training in the use of electronics and simulators. Qualified examiners will give exams geared toward practical factors.

The STCW Code stresses the responsibilities of companies to ensure compliance with the revised standards. *Watchkeeping* is of primary importance to the Convention. Two areas that will receive careful scrutiny include steps a company has taken to assure that its crews can function unimpeded by fatigue, and how the company deals with significant problems such as language barriers between multinational crew members.

Of equal importance is a revised provision that allows a duly authorized Port State officer to conduct a survey of any vessel entering the country's port "if there are clear grounds for believing that (watchkeeping) standards are not being maintained."

Inspection is no longer limited solely to verification of crew numbers and certificates. An accidental or illegal discharge of oil can trigger a more in-depth investigation, detention of the ship, or imposition of economic sanctions by a receiving country. The Flag State issuing the certificate is charged with conducting its own investigation against the violator and enforcing penalties and other disciplinary measures for STCW violations. A significant increase in safer shipping is expected from STCW changes.[23]

THE INTERNATIONAL SAFETY MANAGEMENT CODE

The International Convention for the Safety of Life at Sea (SOLAS), 1974, as amended by 1978 and 1988 Protocols, has been a strong source of protection for the marine world. A significant recent change to this Convention should result in safer shipping practices by oil tankers and bulk carriers.

The United Kingdom's P & I Club analyzed the causes of marine casualties between 1987 and 1992 and reached conclusions almost identical to the USCG Towing Vessel Study. The majority of marine casualties, even where mechanical and equipment failure is involved, are directly linked to human error, and the compounding of such errors by the organizational structure of operating companies.[24]

Based on these findings, IMO conceived the International Safety Management Code (the ISM Code), an innovative approach to regulation of human activity. An amendment to Chapter IX of SOLAS now incorporates this code in its entirety, and mandates compliance for all signatories to that Convention by 1998 for passenger ships, tankers and some other vessels, and by July 1, 2002 for all other ships. The ISM Code:

*"... (establishes) an international standard for the safe manage-
ment and operation of ships by setting rules for the organization of
company management in relation to safety and pollution prevention
and for the implementation of a safety management system."*[25]

GOOD GUY COMPANIES

The ISM Code strengthens the first level of the safety net by ensuring
that operators of oil vessels commit to careful operations *every* time *every*
vessel sails the seas, and communicates this philosophy to all ranks in the
organizations. The ISM Code changes the incentives for conducting *business as usual.*

In the case of *Berman*, schedule was paramount, time was money, and
cost-cutting was the name of the game. The ISM Code now rewards *good
guy* companies. Falling out of compliance may become more costly than
operating by the rules. Under the ISM Code, a company must develop
and have in place by 1998 an approved Safety Management System
which addresses safety and environmental protection policies. All phases are covered:

- How to ensure safe operations.
- Defined lines of authority and communication ashore and on
  board ship.
- Procedures for reporting accidents and safety defects.
- A plan for responding to emergency situations.
- Internal audits and management review.

Operators who comply receive a Document of Compliance and a
Safety Management Certificate, valid for three to five years, subject only
to an annual audit to verify continued compliance. Classification societies
issue the initial certificates, conduct periodic audits and handle renewals.
According to Lloyd's Registry, 20,114 vessels in the world fleet involved
in international trade are now subject to the ISM Code, and more than
11,000 companies operate under it. Proponents of the ISM Code point out
that the benefits to the shipping industry are enormous. Enforcement
alone may reduce the $10 billion a year that the industry is reported to
lose from damage to ships, cargo and humans.[26]

The intended effect of the Code is that non-complying *bad guys* will
simply find their trading sphere shrinking. *Bad guy* shipping companies
increase the chances their ships will be targeted for inspection and
stopped frequently outside port or in the harbor before any business is
transacted. Civil penalties, detention, missed schedules and inability to

operate all serve as substantial negative incentives for a company to move into the "new" arena of safe management. Programs designed to better train and prepare every employee of the company for a more environmentally and safety-oriented operation ultimately become a cost-effective part of doing business.

CDR Sharp of the USCG summed up the shifting dynamics:

> *"A lot of companies fix what breaks to get by and make changes only when they are hit over the head by a two-by-four. The changes the USCG and the ISM Code are creating are the two-by-fours needed in the shipping business."*[27]

INTERNAL AUDITS: OCIMF

In a most encouraging turn of events, the shipping industry is taking action now, not waiting until the ISM Code's active date of 1998. Texaco, Inc. and other oil companies have instituted internal auditing systems. In mid-1989, Texaco established an *environmental* auditing staff to evaluate the status of all facilities and practices to see if they are in compliance with local, state, federal and international laws. Audits provide top management with an honest evaluation of how the facility stands in terms of spill prevention and quick response. Texaco, through hired parties, conducts a regional review of all plants in each region on a three-year basis.

David Davidson, an environmental auditor with Texaco and former USCG member, explains why a major oil company would take such aggressive steps:

> *"While I'd like to say that we do all this for the good of the environment, there are other factors involved. We are much better off finding our own problems before the EPA comes in and shuts us down. We find that our auditing staff is nit- picky to a high degree. When we see what needs to be corrected, we take prompt action. From a consumer point of view, we want to avoid a pollution incident before it occurs. If we have such a casualty, not only is the cost to the company significant, but public opinion is such that people get into the habit of cutting up our credit cards. We don't like to see that happen. A not-insignificant consideration is criminal sanctions for our chief executive and board members. Our principals are not happy about the idea of going to jail."*[28]

Consumers who ever doubt their individual power may keep Mr. Davidson's words in mind: *Cutting up your gasoline credit card works.*

Phillips 66 has initiated an innovative safety program called SCOPE. The company brings in field experts who review their operations and define safety in the work place, monitor performance of employees in the field, and provide positive follow-up for unsafe practices. Employees are encouraged to participate in the program and offer their own feedback.

In addition to SCOPE, Phillips also instituted a self-tracking Substance Release Record (SSR) form to help report oil spills and gas releases to the EPA and local authorities. While "self-reporting" is required by law, Phillips found that SSRs provide instant feedback on how well the company is protecting the environment.

In the year following *Berman*, Phillips experienced a 90 percent *reduction* in reportable events in the first quarter of 1995. This reduction rate equates to a significant cost savings of almost 40 percent less in fines paid from first quarter 1994 to first quarter 1995. Self-control means saving big money for big business – and the environment benefits, too.[29]

Independent oil companies, not just the majors, are addressing the people end of the equation. The Oil Companies International Marine Forum (OCIMF), an organization composed of 34 independent oil companies worldwide, instituted its own policing of shipowners. IMO confirms their importance. Today, major oil companies own less than 9 percent of world tanker tonnage. Independents run the tanker show.[30]

One of OCIMF's more innovative programs is the Ship Inspection Report Programme (SIRE). Realizing that Flag States and classification societies have not been doing their job, OCIMF instituted a system in which their own inspectors perform vessel surveys. Inspections determine whether the ship is in good condition and the crew properly certified to protect man and the environment. Inspection results are entered into a computerized database maintained by OCIMF in London and can be released to participating OCIMF members and approved non-members, such as companies or organizations chartering oil-carrying tankers, bulk oil terminal operators, port and canal authorities, and governmental agencies having Flag or Port State responsibility for tanker safety. OCIMF believes that making reports available to those properly interested enhances tanker safety and reduces the burdens on ship crew and officers caused by duplicate and redundant inspections by the Port State, the charterer, P & I Clubs, and classification societies. [31]

Classification societies are instituting self-policing policies because of an unscrupulous but common practice that cannot be ignored. Shipowners unable to meet operational safeguards and updated regulations may find a dubious classification society to issue them an unearned certificate. The Secretary-General of IMO spoke out against this unsavory practice in his World Maritime Day speech:

"The majority of shipowners accept their responsibilities and conduct their operations with integrity at the highest level. Some

others quite deliberately move their ships to different trading routes of governments in order to avoid stricter inspections and controls; they would rather risk losing the ship and those on board than undertake and pay for the cost of carrying out the repairs they know to be necessary. Some governments are also quite happy to take the fees for registering ships under their flag, but fail to ensure that safety and environmental standards are enforced."[32]

The four largest classification societies in the world, American Bureau of Shipping (ABS), Det Norske Veritas (DNV), Lloyd's Register of Shipping (LRS), and International Association of Classification Societies (IACS) are modifying their transfer-of-class agreements. A shipowner needing repairs can no longer avoid doing the work by changing to a different classification society without others learning about the move. The four big classification societies release information to parties such as governments and insurance companies about any changes in class and the reason for the change. They are instituting higher standards for hiring and evaluating vessel surveyors, and are beginning to suspend member vessels whose surveys are overdue. Actions by the large societies cannot be underestimated. IACS, for example, has the largest membership with some 11 member societies, about 40,000 vessels, and represents 90 percent of the world's merchant tonnage.[33]

Time will tell if new regulations from the international community will whittle away at destructive practices driven by the *bottom line* approach.

MARPOL 73/78

Two types of mariner states control shipowners and vessels, the *Flag State* and the *Port State*. The Flag State is the country from which a ship receives its license and certification that the vessel meets applicable provisions of international convention safety standards and, where appropriate, also meets that country's requirements. The Port State is the state whose port receives the vessel. Each Port State is authorized to verify that the provisions of controlling international conventions are met by examining the foreign flag vessels. If provisions are not met, they may then detain the ship or take whatever action might be necessary to ensure safe operations at sea.

Many regard the control a Port State exercises over ships in its harbors as the final layer in the safety nets of the marine industry. When the shipowner/operator, classification society, Flag States, insurance companies and P & I Clubs fail to do their job of safety inspection and quality assurance, the Port State steps in.[34]

Of all the IMO conventions discussed, the International Convention for the Prevention of Pollution from Ships (MARPOL), 1973 is the most comprehensive pollution instrument. Annex I of MARPOL contains regulations controlling operational discharges of oil from tanker ships and establishes preventative measures to minimize oil pollution in the event of an accident. Other MARPOL annexes address bulk noxious liquid substances, packaging of hazardous substances, sewage and garbage.

MARPOL gives Port States the right to enforce safety on their own waters. In the past, MARPOL inspections ensured that no unsafe conditions existed, but were limited to checking the following:

- Flag State-issued certification documents.
- Shipboard manuals.
- General physical condition of the ship and its equipment.
- Marine pollution compliance.

Again, the emphasis of these previous surveys was on technological compliance, not on prevention through people. Amendments to MARPOL extend Port State control to carry out essential procedures necessary for preventing marine pollution incidents. An in-depth MARPOL examination can go beyond a static structural inspection, and may now include equipment and operational tests.

An inspector in a Port State can prevent a ship from sailing "until it can proceed to sea without presenting an unreasonable threat or harm to the marine environment." Under appropriate conditions and notifications, the Port State can allow the ship to travel to the next port of call when, for example, repair facilities do not exist in the country's own port. In each incident, the Port State notifies the Flag State and possibly the classification society of the deficiencies. These all-important amendments to MARPOL entered into force in March 1996.[35]

POINT SYSTEM

Port State control is increasingly important to the U.S. Approximately 95 percent of the country's passenger cruise trade and cargo import and 75 percent of crude oil pumped ashore involve foreign flag vessels. We depend primarily on others to supply our oil and basic resources.[36]

The previous U.S. system of inspection focused on national fleets, not foreign vessels. In May 1994, the USCG initiated an aggressive program in keeping with the concept of *Prevention through People*.

The USCG maintains a Marine Safety Information System (MSIS) computer database that records the Flag State of a vessel, the owner, the operator, the classification society performing the vessel survey, the age of the ship, and its past safety record. Each vessel is ranked by performance record and the above items. Some Flag States have been identified as having notoriously poor safety records, and certain operators or owners are known for their substandard ships. Classification societies have differing reputations for their strict or loose adherence to safety regulations required by the international conventions. Points are assigned from one to 10.

The USCG assigns priority inspections for foreign flag ships based on the number of accumulated points. A vessel assigned the highest risk receives a Priority I ranking and will always be boarded for inspection by the USCG before entering a U.S. port. Essentially those vessels "suspected of presenting an imminent threat to life, the port, or the environment will be targeted for boarding prior to entry into the port." Those bearing a Priority II designation will be boarded under certain conditions and prior to commencing cargo operations or passenger embarkation. A Priority III-rated vessel *may* be subject to inspection after entry into port. Priority IV ships are rarely boarded.

The USCG knows when a vessel will enter U.S. waters. Twenty-four hours before entering a port, any vessel over 1,600 gross tons must advise the local MSO of its intent to sail into the harbor. At that point, the new procedure kicks in. The MSO checks the MSIS database for the ship's assigned priority, and may inspect accordingly.

When the USCG boards the ship, inspectors conduct a thorough examination that includes an evaluation of the crew's ability to carry out essential shipboard procedures relating to marine pollution prevention. If any safety considerations of serious concern exist, the USCG can conduct a SOLAS intervention and assert its full Port State authority. The intervention may include one of many actions:

- Refusing the tanker entry to the U.S. port.
- Detaining the ship until the unsafe condition is corrected.
- Notifying other countries of the substandard condition so they may also deny access to their port.[37]

The U.S. and other Port States must resort to such aggressive tactics because Flag States and classification societies are not doing their job. According to IMO Secretary-General William O'Neil, "We can claim with justice that we have tried to make shipping safer – but we have to recognize that we have not done enough, and that if we do not do more, safety at sea will get worse."[38]

IMO recognizes that the progress made in reducing oil casualties in the 1980s appears to be shifting in the 1990s. Casualty incidents are on the rise.[39] In too many instances, Flag States issue certifications without insisting on compliance with the convention requirements. Surveyors for classification societies conduct sham surveys. Owners and operators shave costs by delaying repairs on control equipment such as frayed tow lines.

The U.S. and other countries who are victims of increasing marine disasters are finding new ways to say, "No, take your substandard shipping elsewhere." The goal of the point system initiative is simply stated: "The initiative is expected to encourage those responsible for substandard ships to either cease operations in U.S. waters, or adopt management practices that ensure compliance with accepted standards.[40]

With the Europeans, Japanese and South Americans all instigating targeting activities on poorly rated vessels, the hope is that substandard operators will be driven out of U.S., Asian and European waters.

MARGINAL OPERATORS

Before the *Berman* spill, the *Morris J. Berman* barge met all the criteria for a targeted vessel. The owner/operator was a known polluter with a bad prior history who was prohibited from operating in New York waters because of the owner family's infamous environmental record. The management of the parent company and all interconnected businesses ran a less-than-responsible operation.

Will the USCG extend its international point system to domestic U.S. operators like the Berman/Frank family? The U.S. needs to apply the same thinking for domestic shipping as it now appears to focus on foreign vessels entering our ports with a system established and personnel provided, so that the Captain of the Port, like Capt. Ross of the MSO in San Juan, Puerto Rico, can do something about substandard U.S. operators.

Under current regulation and law, the authority of the Captain of the Port is very broad. Based on prior knowledge, the captain can require a U.S. vessel to anchor outside port while the USCG inspects the ship. The Captain of the Port can refuse a vessel entry until he/she deems the ship safe. Once in port, the captain can detain the vessel until safety considerations are met. The Captain of the Port can order a tug assist if one is required. Even with uninspected towing vessels, the tug must meet certain minimum standards. If a violation of any standards is found (oil in the bilges, a faulty steering system, no licenses, fire hazards or a frayed towline) the Captain of the Port can order the tug to wait at the sea buoy outside the port, or detain the vessel from moving until the safety hazard is repaired.

Practical problems prevent exercising this control. Almost without exception, every local MSO Captain of the Port has more jobs to do in a day than he or she can reasonably complete. Rarely can the captain divert much-needed attention from mandatory duties such as routinely scheduled ship inspections, responding to groundings, collisions and the myriad other matters that demand attention 24 hours a day.

Even the new point system contains a major gap by not applying to smaller vessels of less than 500 gross tons. Many such ships sail the Caribbean, for example, and are not subject to international treaty regulations. How can the USCG address this problem? In some districts the MSO has instituted a boarding program similar to that now used for larger vessels. For example, in Miami's Seventh District the USCG boards smaller ships, regarding them as accidents waiting to happen. Whether other MSOs will initiate a similar boarding procedure is an open question.[41]

Only time will tell whether or not Port State controls work, but the maritime community believes that this system hits companies where it hurts, in their economic belly. Ships entering foreign ports are on tight schedules. When they are subjected to strict enforcement of convention standards by lengthy inspections, schedule delays occur. Failure to comply with engineering concerns and operational requirements may result in even more harsh economic sanctions, such as detention at port or denied entry. Under Port State control, a vessel could face the ultimate economic control: being prohibited from conducting its trade in certain ports and regions of the world. Given the choice between not trading or compliance, many substandard vessels and operators may be forced to conduct safer operations or go out of business.

While the international community seems to be heading in the right direction and much good work is being done, there are still enormous impediments to the ability of poorer developing countries to exercise effective Port State control of ships entering their harbors. Puerto Rico's experience with *Berman* is a perfect example.

ISLAND POLITICS

Immediately after the *Berman* grounding, Sen. Freddie Valentin Acevedo, the president of Puerto Rico's Senate and chair of the Natural Resources Commission, introduced a resolution giving the commonwealth far-reaching control over its own waters. The proposed law, Ley 622, sought among other matters to empower Puerto Rican officials to inspect all foreign flag and U.S. vessels carrying oil or hazardous substances upon entering or leaving its harbors at least once a year.

Additionally, the resolution proposed that ships would remain at least five miles off the coast until the final approach to or exit from the island, so as to avoid the difficult stretch of reefs along the coastline; required that any ship using Puerto Rican ports be a double hull (not waiting until the federal phase-in time of the year 2015); and legislated absolute financial responsibility for any damages caused by an oil spill.

This daring piece of legerdemain was introduced during the highly emotional months following the spill and after extensive hearings held by the legislature in which various agencies submitted reports and testi-

246

fied about the tragedy happening on the island. The full extent of the cleanup was still unknown, but looked ominous. The Puerto Rican community was agitated and disturbed by the sight and smell of the hideous mass of oil smothering its beaches and shores. The uncertain future of Puerto Rican waters, given the island's proximity to the world's oil trade routes, demanded innovative action.

Unfortunately, Sen. Valentin's bill fell prey to Puerto Rican politics, Machiavellian at best, and to economic realities. After passage in the legislature, Gov. Rossello of Puerto Rico received the bill for his review and signature. Before signing, nagging doubts about the ability of his small island to regulate U.S. waters and other concerns caused him to send the bill on to the office of the attorney general in Puerto Rico for comment.

Usually a project of such magnitude circulates in draft to all concerned agencies and individuals before being submitted for review by the legislature. Those involved in this case included the Environmental Quality Board, the Port Authority, the Power Authority PREPA, the Attorney General's office, and others. By following this unwritten procedure, the powers that be have a chance to add their opinion and the proposal stands a much better chance of clearing political hurdles. Those in charge provide necessary input, and no agency feels left out of the process. No one is surprised or placed in the uncomfortable position of giving advice without advance warning. Valentin's resolution met an unfortunate demise.

The appropriate attorneys conducted their own study. They decided to pan the bill based on two significant drawbacks. Right or wrong, the legal

Sen. Freddie Valentin Acevedo with Puerto Rican fishermen following the *Berman* spill. (Reprinted by permission of Sen. Valentin.)

advisors concluded that Puerto Rico cannot usurp federal rights and preempt the USCG's authority to inspect and control ships coming into the commonwealth's ports. Secondly, the responsible attorneys in the Attorney General's office consulted with the USCG and found, to their dismay, that Puerto Rico did not begin to possess the administrative capacity necessary to inspect all foreign flag vessels entering its ports.

The general consensus was that the Commonwealth Port Authority was barely able to manage the limited matters within its current jurisdiction. With unskilled inspectors and no money to train them properly, the Valentin legislation added another level to an already overwhelmed agency that resulted in an untenable situation.[42]

Capt. Ross agreed:

> "The commonwealth doesn't have the depth to take over inspections of foreign vessels. As much as I applaud the intention of the Puerto Rican government, the cost and time it would take to train unskilled personnel to conduct Port State inspections would be prohibitive. You need a marine surveyor to perform a competent inspection of a tanker or other vessel. The USCG spends four years of resident and on-the-job training to turn our people into qualified marine surveyors, and we're starting with an already trained USCG member. Can you imagine the cost to the Puerto Rican government and the time involved to train an untutored staff to take on the job the USCG is already performing in San Juan?"[43]

The governor vetoed Sen. Valentin's bill. This same scenario may well play out in other Caribbean nations where trained surveyors and money to fund inspections is lacking, unless underdeveloped countries receive strong financial and technical backing from IMO, the World Bank and others.

Potentially, Port State control could end substandard shipping in the regions that protect their own waters. The world is far from reaching this goal. The consensus is that Port State control is yet another valuable device in the international arsenal of strategies to reduce, limit and eliminate the spheres of operation for shoddy operators. Yet, the current realities of inexperienced staff and limited funding make this an "impotent" tool.

INTERVENTION ON THE HIGH SEAS: LEGAL PIRACY

Piracy is as old as shipping itself. A more powerful entity can forcibly board a ship and take charge of its cargo. Even the honorable international community recognizes that sometimes piracy is necessary. In 1969, states adopted the International Convention Relating to Intervention on the High Seas in Cases of Oil Pollution Casualties (Intervention Convention). The convention became effective in 1975. This treaty gave coastal states the right to take action to "prevent, mitigate or eliminate grave and imminent danger to their coastline or related interest from pollution or threat of pollution ... following a maritime casualty." Such action may include boarding a ship, salvaging it or sinking the vessel. Initially limited to oil, the provisions now extend to other substances.

France used the Intervention Convention when an Iranian tanker carrying 220,000 tons of oil lost total propulsion in mid-channel near Cherbourg. The captain wanted to propel his boat through the Straits of Dover using tug power only, a very risky venture given that the straits are no more than a mile wide. The French refused permission. After consulting with the British, the French moved the vessel to a quiet part of their coastline and then lightered its product from the ship's hold to another vessel until power was restored.

Under the Intervention Convention, a state may take "any measure" in dealing with a vessel, so long as the measure is commensurate with the dangers involved in the incident. An intervening state can go all the way in its response. In 1981, the French bombed a ship out of the water when its refined products threatened the Corsican coastline. The bombing produced "in situ" fire and literally burned away the oil.[44]

Perhaps when all else fails – internal management systems, inspections by Flag and Port States, training and licensing programs – Port States will be forced to resort to the last arena of prevention: legal piracy. That legal piracy has become an accepted practice is evidence of how serious the overall management and control of shipping has become. During this uncertain period in which new regulations are being enacted, unscrupulous shippers are still finding ways around older regulations, and the world tanker fleet continues to age at an alarming rate, some nations are stepping in to take the ultimate control.

PROGRESS IN THE FIELD

BACK TO THE BASICS

Ours is a small world, growing smaller every day. The welfare of our reefs is affected by waters surrounding non-U.S. coral reefs. "The health of one country's marine environment ... is frequently dependent on the actions of neighboring countries."[1]

Pollution on land flows into the sea. Trees cut in an old-growth forest, mangroves removed from coastal shores and farming on upland slopes result in soil erosion and losing natural siltation sinks. The coastal zone is no longer limited to the area immediately bordering the water. All land masses are part of the ocean fringe.

If we do not shift our management policies and attitudes toward protecting the coastal zone and the oceans touched by this extensive area, coral reefs will undergo further, possibly irreversible, degradation. According to experts, "It has been predicted that more than two-thirds of the world's coral reefs will be considerably degraded within 20 to 40 years unless effective management of the resources is implemented urgently."[2]

Managers worldwide are changing their attitudes regarding coastal zone and marine habitat protection. Much like the shipping industry, the conservation community is now including people in its equation.

At one time conservation meant protection, biological inventories, and fencing in biologically significant areas. Conservation actions then were simpler, more direct – and rarely successful. According to environmentalist Evelyn Wilcox:

> *"People, along with their needs, wants and aspirations, were not part of the equation ... what a Pandora's box was opened when we recognized that successful conservation entails active involvement of the social, political and economic sciences. Programs and projects receiving support today invariably include public participation, grass-roots liaison with government, human needs assessments, policy development and many other non-biological aspects."*[3]

Biologists and researchers working in the field recognize that successful protection of the environment requires the involvement of the people who live, work and depend on the very ecosystems at stake. Just as IMO and the world community are partnering with industry, shipowners and crews to create a safer maritime system, conservationists at all levels are bringing local citizens into the picture.

Conservation of the marine environment faces major hurdles because of certain unavoidable realities. Coral reefs are found in more than 100 countries. With the limited exceptions of some industrialized nations, most reefs border developing countries. These coastal communities are often home to the poorest and least-educated among the rural poor. Given increasing population shifts of impoverished and sea-dependant masses toward the coastal zones, where more than 60 percent of the earth's 5.6 billion inhabitants currently live, this demographic is unlikely to improve. Haiti is an extreme example of this modern problem. Overpopulation, poor agricultural techniques, and depletion of commercial fish stock in Haitian reefs has created a nation of "environmental refugees." Hungry Haitians can no longer turn to their once-rich natural resources to feed themselves.[4]

Poverty demands action, but we often take the shortest and easiest route to problem-solving. The long-term approach is the exception rather than the rule, and "starving people are understandably immune to suggestions that they practice resource management."[5] An evolving cycle of governmental or local funding for unsustainable projects creates jobs and ready sources of cash. *Ecotourism,* which draws tourists to an area to appreciate its natural wonders, provides some solutions, but may conflict with other incompatible factors: lack of community participation; inadequate infrastructure to support tourists who require beds, meals, transportation and equipment to enjoy the resources; *outsider* companies who profit from tourism and disregard the local work force; over-exploitation of already stressed resources; and more.

There is light at the end of this tunnel of poverty, short-term thinking, and coastal degradation. Those working in marine conservation are framing solutions to this host of problems.

From studies conducted by Worldwide Fund for Nature (WWF), the Rosenstiel School of Marine and Atmospheric Science, IUCN, the U.S. Agency for International Development (USAID) and the International Coral Reef Initiative (ICRI), scientists and managers in the field are seeing common threads emerge. Successful marine conservation requires a multi-pronged attack in each of these five major areas:

- *Education:* Adequate data to evaluate the condition of the world's reefs and create conservationists from local communities.
- *Poverty:* To better understand and address "marine-related poverty."

- *Link Between Land and Sea:* Link decisions made on land with their effects on the sea; use scientific information to aid managers and encourage local populations to preserve their resources.
- *Time:* Look to reasonable time frames of at least 10 years or longer for success.
- *Funding:* Projects require long-term funding of 10-year commitments from donors, project leaders, beneficiaries, governments and NGOs.[6]

POLITICAL FRAMEWORK

No marine park is an *island*. No longer can we designate a local area as a protected zone and ignore the people who live on or near its boundaries. The days of isolated conservation efforts are numbered, just as the days of the lone, unregulated seafarer are gone. The most workable approach lies in tailoring umbrella programs to local conservation efforts. These programs range from far-reaching U.N.-sponsored projects to regional monitoring networks. At the end of this chapter is a brief summary of some conservation programs, but certainly not all, in which common threads recur and point to successful coral reef preservation.

YEAR OF THE REEF AND CORAL REEF INITIATIVES

1996 has been designated as the International Year of the Reef after an extensive study of reef degradation found at 28 locations worldwide. More than 120 scientists of the University of Miami's Rosenstiel School of Marine and Atmospheric Science concluded that educating local citizens living near the world's reef is a high priority. Once residents realize the many benefits reefs provide, they can become better reef protectors. International collaboration between scientists, administrators of reef parks and reserves, and environmental groups is essential for reef protection. The International Year of the Reef designation emphasizes the initiative's goals.[7]

Shortly after the U.S. ratification of the Biodiversity Convention, the U.S. joined seven other countries to launch the International Coral Reef Initiative in December 1994. ICRI's goal is to mobilize "political will" at local, national, regional and international levels for "on-the-ground practical cooperation."

Because of its focus, the Initiative serves as a useful metaphor for the work that other conventions, NGO initiatives and governmental models seek to achieve. ICRI has three major objectives:

1. *Conceptual Management:* To promote sustainable management by encouraging governments to develop national coastal reef initia-

tives and to incorporate those initiatives into their existing development plans.

ICRI calls for integration of coastal zone management plans with land use and urban development planning. Sustainable use of marine ecosystems is the goal. An integrated management system translates into strengthening coastal and marine ecosystem management, bringing the private and public sector into the planning process as willing participants, and sharing management practices between countries.

The Framework for Action asks governments to:

- Act to develop coastal management plans.
- Integrate sound land-use practices with marine protection.
- Plan for sustainable fishing, tourism and enforcement.
- Integrate local and regional plans with international treaty pro tection.
- Develop legislation where needed.
- Improve the overall socioeconomic conditions of local communities by teaching citizens to benefit from sustainable use of deminishing resources.[8]

2. *Capacity Building:* To strengthen national and local capacity to implement these partnerships.

This two-step process first raises the awareness level of residents to appreciate the importance of reefs, mangroves and seagrass beds. The second part involves coordination between local, national and regional communities for effective management and sustainable use of coral reefs and their related environments.

Education and information exchange are the keys. Conservation begins with knowledge. Trained data collectors create a better understanding of a country's resources. Taking the information in hand, managers with the requisite expertise can plan resource control. Educating the local population about the value of their resources, making conservation a priority goal, sustainably using dwindling resources, legislating protective laws, and developing private/public partnerships to fund and undertake projects comprise the heart and soul of capacity building. An example of capacity building is the University of Rhode Island's Coastal Zone Management Program at sites throughout the world. Using USAID funds, managers, graduate and undergraduate students from various countries are trained to return to their countries and set up marine reserves.[9]

3. *Monitoring and Information Exchange:* To establish contacts among international, regional, and national research and monitoring programs and coordinate efforts to ensure efficient use of resources.[10]

WIDGETS ON REEFS

ICRI seeks to put *widgets* on reefs and build a global network of coral reef research and monitoring.

Coral reefs are ideal indicators of climate change and the health of the world environment in general. They may provide an early warning about global warming. We know little about the health of today's reefs, although corals have existed for hundreds of millions of years. We do know that understanding long-term trends in reef condition is vitally important to a better grasp of what is happening to our natural world.[11]

Monitoring is essential. "The repeated surveying of organisms or envi ronmental parameters over time helps us understand a variety of natural processes," such as the abundance of certain types of corals, species diversity at a site, the condition of particular habitats, and changes in the environment caused by human behavior or natural processes. Monitoring allows us to see relative changes in a particular reef over time.

Reliable baseline data provides a greater understanding of natural systems and assists in making responsible management decisions. Such decisions might include requiring developers to monitor reef conditions and take corrective actions when sediment from disturbed earth in a development flows into the ocean. A ban on spear fishing, limiting capture of lobsters or other marine creatures, or prohibiting fishing during spawning season for certain species might be called for when stocks are seriously depleted.[12]

Unfortunately, there are few well-trained scientists or concerned individuals able to conduct reliable monitoring. Carefully maintained sites are limited throughout the world. ICRI projects such as the Reef Assessment Program and an ambitious international project called Global Ocean Observing System (GOOS) seek to address this lack.[13]

Another major byproduct of monitoring is accurately mapping reefs around the world. Today there is no composite map of global reefs. Overcoming this visual barrier by placing reefs on the map makes them real for people, where they can become focal points for protection.[14]

ICRI is not simply another bureaucracy. Instead, it works through already established agencies and funding sources. The Secretariat is set to be a short term of three years, the first year based in the U.S. and the last two in Australia. After countries create their own national and regional plans, funding resources within these countries, with help from USAID, should result in a functioning yet decentralized program. USAID provides $90 million a year to developing countries.[15]

Regarding ICRI's likelihood of reaching its goals, U.S. Under-Secretary Tim Wirth said, "It is difficult trying to overcome our differences and work side-by-side, not only on behalf of our own citizens, but the citizens of the world. To those who question whether we can afford to try, we say that we cannot afford not to."[16]

CONSERVATION PROGRAMS

AGENDA 21

The U.N. Conference on Environment and Development in Rio de Janeiro in 1992 produced Agenda 21, the base working document for the future of worldwide environmental protection. Chapter 17 of Agenda 21 calls for new approaches to the sustainable use and conservation of marine living resources under each participant nation's jurisdiction, through management and development programs at all levels. Chapter 17 highlights the need for baseline data; proposes conducting studies to determine whether current utilization of these resources is sustainable; directs integration of coral reefs into land use planning and urban development; and requires periodic review of implementation success by the Commission on Sustainable Development.

WORLD HERITAGE CONVENTION

Enacted in 1972, the Convention for the Protection of the World Cultural and Natural Heritage (World Heritage Convention) is sponsored by UNESCO with participation by 120 countries. The Convention's intention is to identify areas worthy of world heritage status, to protect those sites, provide information and public education about them, and to use heritage locations for research and environmental training. To date, 358 international sites have been designated for their *outstanding universal values*. Coral reefs selected as World Heritage sites include the Great Barrier Reef, Sian Ka'an on Mexico's Yucatan Peninsula, Henderson Island in the Pacific, the Galapagos Islands, and the Sierra Nevada de Santa Marta off the coast of Colombia. Recommended for such status are the Chagos Archipelago, a remote chain of reefs in the center of the Indian Ocean, Belize's Barrier Reef, and Belau.[17]

An example of this program is Puerto Rico's enactment of Law 150. U.S. states, commonwealth and territories have passed the legislation necessary to implement this convention. Significant enactment of this international treaty by Puerto Rico's Law 150 mandates the creation and conservation of inventory sites with high priority because of their biodiversity or the significance of their natural resources, including examples found in special ecosystems, coastal lands bordering coral reefs, mangroves, rainforests, wetlands and marshes. The commonwealth is authorized to acquire, restore and manage this identified *inventory* based on five criteria:

- Areas important to endangered species.
- Natural communities.
- Sites utilized by migratory species.
- Wetlands systems.
- Systems having a special design.

Puerto Rico's Law 150 authorizes review of proposed projects if they threaten a natural heritage site. Puerto Rico is also part of the newly established international Network of Natural Heritage Program and Conservation Centers in Latin America, the Caribbean and the U.S. that help identify sensitive areas to protect species living in those habitats.[18]

STEWARDSHIP - REGIONAL SEAS PROGRAMME

Formed in 1972 by the Governing Council of the United Nations Environment Programme (UNEP), the Regional Seas Programme helps developing countries form Action Plans on a regional basis to protect marine resources from pollution and over-exploitation. The Programme also emphasizes cooperation across jurisdictional boundaries, calls for linking management issues between nations in a given region, and declares national and international communities stewards of the world's seas. Fourteen Regional Seas Projects exist today. The Caribbean Coastal Marine Productivity Program (CARICOMP) resulted from this program as a regional scientific project to study land-sea interaction processes with special attention on the health and status of coral reefs.[19]

MAN AND THE BIOSPHERE

The U.N.'s Educational, Scientific and Cultural Organization (UNESCO) produced the Man and the Biosphere Programme in 1983 to create special resource areas known as *biosphere reserves*. These are internationally recognized highly protected core, areas surrounded by a buffer zone and a transition zone. The core area is for conservation purposes only, while the buffer zone is for research and non-extractive activities, such as recreation, tourism, education and traditional uses. Sustainable development occurs in the transition area if it enhances core area conservation, involves the local community and is managed across boundaries of neighboring countries.

Marine protected areas are defined as "any area of intertidal or subtidal terrain, together with its overlying waters and associated flora, fauna, historical and cultural features, which have been reserved by legislation to protect part or all of the enclosed environment."[20] *Less than 1 percent of the marine area that covers more than two and a half times the total land masses of the world is under protection.* Only 430 Marine Protected Areas have been proclaimed by 69 nations, with another 298 areas proposed as of 1985.[21]

Arnavon, one of the Solomon Islands, has more than 40 species of corals in its reef systems. In 1989, representatives from diverse interest groups in Arnavon met, developed partnerships, and formed a framework for establishing a marine protected area. Fourteen years later, in 1995, Arnavon was designated the first marine protected area in the Solomon Islands. Making this conservation concept work for Arnavon was a tall order. Arnavon is home to three distinct communities with entirely separate backgrounds, two of which have a long history of warfare. A locally-based NGO, The Nature Conservancy, worked with the Arnavon government to overcome hostilities between these culturally distinct groups. The successful outcome in Arnavon was due in large part to involvement of all local citizens. Members met with trained managers to develop plans for conservation of their own resources. Patience paid off. Conservation efforts take time – in this case, "Pacific Time," or "slow time."[22]

Another success took place in the Caribbean at the Les Arcadins Marine Park and Fisheries Project in Haiti. Ignoring the terrible political problems of this small country, Haiti's 1500 kilometers of coastline constitute one of the longest and least environmentally disturbed areas in the Caribbean. Until recently, Haiti's reefs escaped the slow but inexorable urban spread into the coastal zone. Poverty often goes hand-in-hand with population growth. These two factors inevitably led to deforestation, erosion, pollution and over-fishing.

Established in 1961, the World Wildlife Fund (WWF) is a nongovernmental organization with more than 1 million members. WWF began as a small grants maker; today, WWF is one of the largest NGOs and conservation organizations worldwide, with projects across the globe.

WWF targeted a small village 30 miles from the capital of Port-au-Prince and near three offshore islands, Les Arcadins, to establish a marine protected area. Three scientists, primarily from Puerto Rico, undertook an extensive survey of the proposed park area to obtain the necessary data to map the park. Fishermen in the village of Luly were full participants in the project from start to finish. A fish-breeding reserve, off-limits to fishermen and tourists, helps to regenerate a depleted fish population. Local fishermen monitor the reserve. Other locals are learning how to explain marine conservation to the residents.

WWF targeting the community with an aggressive education program, teaching useful agricultural practices to residents and encouraging them to engage in sustainable farming rather than depleting fishing stocks. WWF translated scientific information into easily readable Creole language pamphlets for use in schools, started children's summer camps, conducted community meetings, and established fishing teams.[23]

ANCIENT TABOO: RETURN TO TRADITION

Palau is an example of a special marine ecosystem which benefits from a partnership of local community and NGO involvement. Located between the Philippine Sea and the Pacific Ocean, Palau consists of 340 islands along a 400-mile chain. Considered one of the *Seven Underwater Wonders of the World*, Palau has more than 625 corals and 1,400 fish species inhabiting its extensive reef system. For 200 years, residents have lived in harmony with the sea, exercising ancient conservation rites to protect their fishing stock and reefs – until the 20th century intruded.

More than 40,000 tourists visit Palau annually. Power boats allow fishermen to reach formerly far distant reefs. Ice chests and refrigerators provide extra storage for larger catches. A multitude of different hooks, spearguns and other devices pull greater numbers of fish from the sea. A government patterned after the U.S. model replaced the tribal chiefs of each village. Soil erosion, over-fishing, destruction and pollution of mangroves and subsequent decay of the reef system resulted in two small fishing villages taking matters into their own hands. Their chiefs reinstated an ancient taboo, the *bul*, a traditional conservation method which prohibits fishing between April and July, the major spawning season for many species of local fish.

The Nature Conservancy helped with this process by preparing the natives of Palau for the inevitable onslaught of 21st century tourists. Organized in 1951 as an NGO, The Nature Conservancy's mission is "to preserve plants, animals and natural communities representing the diversity of life on Earth by protecting the lands and waters they need to survive."

Under its Micronesia Program, The Nature Conservancy opened a field office in Palau and initiated programs with locals, many of which are education-based. Their efforts are paying off. In one village, the chiefs are proposing a year of rest for the reef, with monitoring to determine the effects of their modified *bul*. The Nature Conservancy teaches residents how to conduct monitoring, funds basic biology training, surveys beaches for sea turtle poachers, trains locals in the use of mooring buoys for diving boats to protect the reefs from anchor damage, and offers fishermen an alternative occupation: guiding sport fishermen in a responsible fashion.

A new sea farming industry is being created to prevent depletion of fish stocks within the next five to 10 years. Giant clams are cultivated for food and export. Islanders visit a clam-growing facility for a three-month internship and learn to cultivate these creatures. At the end of their internship, each intern receives 1,000 pieces of giant clam to plant in the reefs closest to their home village. In this way, reef biodiversity is maintained and a new cash crop finances the Palauan economy.[24]

NATIONAL MARINE SANCTUARIES

In 1975, the U.S. Congress designated the first two marine sanctuaries: a coral reef off Key Largo, and the Monitor, an area off North Carolina where a civil war ship sank. Twelve other sites including Pennekamp Park in the Key Largo National Marine Sanctuary have since been added to the list.

Pennekamp is one of the most popular diving and snorkeling areas in the U.S., with almost two million visits annually.[25] Because of its popularity, Pennekamp is slowly deteriorating. As one diver stated, "We will kill the coral reefs if we're not careful, by ignoring their silent plight and loving them to death."[26]

One of the greatest resources for saving this fragile reef system may come from those who benefit the most from its existence: the divers, dive shop owners, and associated retailers who service the hundreds of thousands of visitors to the reef. These commercial entrepreneurs acknowledge the negative impacts of their users and the decline of fish and coral. Many have initiated programs to educate divers and snorkelers about the harm caused by touching, removing or striking coral. Divers employed by the shops jump into the water and flash signs advising caution when overeager tourists appear poised to pounce on fragile coral structures. Self-interest motivates education and preservation.[27]

RAMSAR RECOGNITION

The Sixth Conference of Contracting Parties to the Ramsar Convention 1996 called for the conservation and wise use of coral reefs. Written and signed in 1971 in Ramsar, Iran, Ramsar is a highly regarded international treaty initially conceived to protect wetlands globally. Until the Sixth Conference of Contracting Parties, its primary focus was land based wetlands, although the treaty includes coral reefs in its definition of wetlands. In March 1996, more than 90 countries party to the convention recommended designating coral reefs and their related ecosystems for future Ramsar sites and urged conservation of reefs globally. While non-binding in nature, Ramsar sites are subject to national and international protection. The 1996 recommendation ensures that coral reefs will be included as future new Ramsar sites.[28]

WESTERN HEMISPHERE PROGRAM

In 1973, the U.S. Fish and Wildlife Service introduced the Endangered Species Act (16 U.S.C. Sections 1531-1543) which gave new meaning to the 1941 Convention on Nature Protection and Wildlife Preservation in the Western Hemisphere (WHC).

The WHC establishes parks for public education, nature reserves for plant and animal protection, wilderness areas closed to motorized vehicles, a migratory bird management program, and regulates permits for the export, import and transit of protected flora and fauna.

WHC's goal is to help countries achieve self-sufficiency in managing and conserving their own resources. Well-trained conservation and natural resource management specialists in Latin America and the Caribbean apply a multifaceted approach, instructing participants through long-term education programs at the graduate level, short-term training such as local workshops, and instituting a small grants program to provide seed funding to fledgling conservation groups.[29]

STEWARDSHIP

Biological oceanographer Gilberto Cintron-Molero, a renowned biologist with extensive experience with coral reefs and coastal zone management, shared his insights into the successes and failures of WHC's small grants program. His comments are instructive for future victories in the field:

> "What we are most concerned with in this program is capacity building, not necessarily good science. Science is important, but more important is environmental awareness. In order to protect the reefs, we need to build a constituency in the various countries from the people who live and work there. The way we've found to do this is through our Small Grants Program.
>
> "How this plan works is that we find local grassroots groups, conservation organizations and NGOs. The USFWS makes the entity a small grant of several thousand dollars ($1,000 - $2,000) to help them conduct some community type of environmental activity. In this way we develop partnerships with locals and linkages with citizens from the ground up. USFWS sends our people there to conduct training. With a small initial capital contribution and staff help, we bring in partners and get the most for our money.
>
> "What the WHC program does so effectively is to train people who have a high degree of education, for example, an MA level degree, but are in need of resource management skills. We provide technical level programs for protected area managers and more in-depth training on a longer-term basis. The demand for such qualified individuals is so great that I don't know of any who are unemployed."[30]

As with other successful ventures, the WHC approach stresses what works best: a *bottom-up* approach within communities surrounding the protected area or project, and *partnerships* with local citizens, local governments, area conservation groups and funding NGOs.

THE GREAT BARRIER REEF MODEL

ASSESSMENT, COMMUNICATION, EDUCATION AND COMMITMENT

How do we begin to protect coral reefs in a world that traditionally has never conceived of the need for such protection? One of the best answers that comes from the field is *ACEC*: Assessment, Communication, Education and Commitment.

Assessment of the reef's condition: Study and understand the impacts of man and nature on both heavily-populated and remote reef systems.

Communication of information: Those charged with managing reefs must communicate with scientists, environmentalists, NGOs, governments and students about how best to manage the environments in and around coral reefs, including links to the land.

Education: Promote sustainable reef use and conservation practices among the user public and local citizens who live near the world's reefs.

Commitment: Commit to long-term monitoring and marine conservation projects and provide the funds and time needed to accomplish them inception to completion.[1]

THE GOOD, THE BAD AND THE UGLY

The small island of Puerto Rico is an example of *the good, the bad and the ugly* to be found in each of the four areas of ACEC.

According to longtime residents, Puerto Rico was a paradise not so many years ago. Like many other Caribbean islands, Puerto Rico was heavily wooded before the Europeans invaded and cleared away forests for sugar cane plantations. By removing the permanent forest cover and replacing it with a seasonal, impermanent layer of sugar cane, the land lost its natural self-regulating ability to hold nutrients and sediment. Sugar cane farming eventually gave way to even poorer agricultural practices, followed by shopping centers and coastal development consisting of wall-to-wall housing, big tourist hotels, and lots of cement.[2]

Because Puerto Ricans are heavily dependent on the sea, over-fishing and exploitative fishing methods became standard practice that led to physical destruction of the reefs and depletion of mature and, eventually, juvenile fish stocks. Sitting at the heart of one of the world's oil choke points, Puerto Rico experienced repeated oil spills, with *Berman* being the most recent. Until recently, coastal zone management has been virtually nonexistent.

The population is plagued by typical signs of poverty: limited employment opportunities, ignorance of national and environmental treasures, crowded housing on and near the coastline, and an unsustainable tourist trade that annually disgorges hundreds of thousands of visitors into an already overcrowded central area.

The end result spells bad news for the island's reef ecosystems, and the consequences have been downright ugly. Except for a few coral patches scattered along Puerto Rico's shores and nearby islands, biologist Carlos Goenega found serious degradation of almost all coral reef systems surrounding the mainland. The most heavily impacted reefs sit at or near openings from rivers where silt from denuded soil flows directly onto coral structures. Human waste and garbage pour directly into the sea, making almost 45 percent of the island's waters unhealthy for reefs and unsafe for recreational swimming.[3]

EDUCATION IN ACTION

Educating the Puerto Rican public is one useful method of combating this massive reef degradation. Aileen T. Velazco-Dominguez is coordinator for the Program of Aquatic Resources Education sponsored by Puerto Rico's Dept. of Natural and Environment Resources (DNR). Velazco's program attempts to change old attitudes and perceptions about the reefs without causing major social upheaval.

Velazco's fisheries project combines the locals' love of fishing with lessons on the marine environment. Along with practical instruction, DNR employees explain to locals how the fish, reefs, seagrass beds and mangroves all support each other. Participants learn what practices harm the reef, and how the overall health of all parts of the larger ecosystem affect the future fish catch. This fishing education program is so popular that more people enroll each weekend than the project can handle.

DNR also established a training course for teachers, many of whom know little about their own environment, so they can better educate their students. DNR translated well-established teaching programs such as "Project Wet" and "Aquatic Wild" into Spanish for classroom use. Training courses encourage a hands-on approach by giving students the opportunity to learn and study in the field.

La Nueva Dia, the largest Spanish newspaper on the island, publishes *Pesca Recreativa* ("recreational fishing"), a DNR supplement about the aquatic environment that reaches 260,000 people and stresses wise husbandry of the people's natural resources. The June 1994 supplement included a discussion on the effects of the *Morris J. Berman* spill and how such a disaster emphasizes the importance of protecting the island's natural resources.[4]

Other agencies stress education. Jorge Saliva of the USFWS conducts a public outreach program, talking with schoolchidlren of all ages. Saliva presents visual aids and materials about the environment to explain the importance of protecting the habitat of endangered species. The program explains in an understandable format the reasons for the decline of important plant and animal species and what students can do to help make a difference and reverse the trend. USFWS sets up exhibits at various malls that stress the importance of protecting wildlife. One exhibit showed a photograph of a man killing a sea turtle for its shell, with a representative who explained why sea turtles should be spared. Saliva believes the program is successful:

> *"I believe [it works], particularly where children are the ones we speak with. Kids grasp the idea better than adults. What we teach them, they can teach their parents. For example, take trash handling. Ten years ago, islanders started to get the picture that throwing trash on the beach was a bad idea. Today we really emphasize that point with the students. They are the ones you hear lecturing their parents not to throw things on the beach, in streams, in their own backyards."[5]*

Newscaster Susan Soltero of *TeleOnce* (Channel 11 News) uses her popularity and recognition to visit school systems with her video program about Moises, the celebrity manatee, and educate students about protecting endangered species. Soltero delivers her message in a way that school children can understand. She explains:

> *"After showing them pictures of the different birds, fish and animals living in Puerto Rico that are disappearing from our island, I talk with the children about why this is happening. I use an example that they can relate to, like throwing a diaper into the ocean. I ask them, would you like to be swimming along and run into a foul, smelly diaper floating in the water? They usually groan at that idea. Then I say, what do you think it's like for a sea turtle or a manatee to swim into a piece of leftover plastic diaper? They get the point very quickly."[6]*

In her text, *Lessons from the Field*, Evelyn Wilcox highlighted the importance of education in marine conservation:

> *"Judging from the responses on the issue of training and education, project leaders were convinced that training in sustainable use and local community public awareness programs will make a positive difference in the manner in which natural resources are being used (97 percent). Respondents believed that publicly aware local*

communities are the best insurance for long-term conservation of marine resources (87 percent)."[7]

COMMUNICATION AND COMMITMENT: INTEGRATED COASTAL ZONE MANAGEMENT

Puerto Rico bit the bullet and revised its outdated 1975 coastal management code to stress the importance of links between land and sea. After carefully studying the coastal zone area on an island the size of Puerto Rico, drafters opted to designate their entire island as a coastal zone. Puerto Rico's new code requires that any federally funded project in the coastal zone must be consistent with the plan.

The planning board is charged with the final review of any submission, although the Puerto Rican Coastal Zone Management office of DNR acts in an advisory capacity for the board. DNR reviews each project to see what direct or secondary impacts the proposed construction will have on the entire watershed, from wetlands to the coastal shore, and finally to the ocean. Developers will be required to revise plans and mitigate negative impacts.

While DNR hopes for funding to update its coral inventory, last completed by Carlos Goenega in 1979, there appears to be no money in sight in the near future. The intent is to map all coral reefs, describe their status, determine what can and cannot be saved, and plan for their future protection by revisions to the Coastal Zone Management Plan. DNR also has a Sensitivity Atlas for Oil Spills, identifying areas most sensitive to oil spill effects and most likely to be impacted, which needs the same type of updating.

For reef protection, DNR coordinated with USFWS to start the mooring buoy program, or *boyas de anclaje*. DNR places buoys near fragile reefs throughout the island and outer island waterways. They also circulate English and Spanish brochures telling boaters what coral reefs are and how boat anchors damage their fragile under structure. The pamphlet instructs boaters how to properly use anchors, identifies where such buoys are located in busy and protected areas, and explains how swimmers and snorklers should be careful when swimming near coral.[8]

WORLD-CLASS SYSTEM: THE GREAT BARRIER REEF

While Puerto Rico's marine protection systems are in their infancy, Australia's management of the Great Barrier Reef (GBR) is the leading model of effective management and protection of one of the world's greatest reef resources. The GBR stretches more than 2,000 kilometers along Australia's eastern (Queensland) coast, covers approximately 348,700 square kilometers, contains 2,600 individual reefs and 300 reef islands (or cays), and supports the greatest diversity of plant and animal life found anywhere in the world: 1,500 species of fish, more than 300 species of hard coral, over 4,000 species of mollusk, and at least 252 species of birds. The GBR is the world's largest coral system; 93 percent of the region falls under the Great Barrier Reef Marine Park, controlled by the Great Barrier Reef Marine Park Authority (GBRMPA).[9]

Tourism in the reef and coastal areas represents a A$1 billion (A=Austrailian dollar) a year industry. Diving or snorkeling the GBR is a marine lover's dream-come-true, where dense schools of fish and abundant marine life surround visitors in canyons of coral. Commercial and recreational fishing pulls in another A$.4 billion annually.

Australia first recognized the need to protect the GBR in the early 1970s, when certain mining and oil interests announced their intent to mine the reef and drill for oil and minerals. Public reaction led to establishing the Great Barrier Reef Marine Park (GBR Marine Park) under the Great Barrier Marine Park Act, 1975. Oil drilling and production were subsequently banned in the GBR region and, by later amendment, control of the GBRMP was placed in the GBRMPA.

Australia is the fifth-largest shipping destination in the world in terms of tonnes of cargo shipped and kilometers traveled, and depends almost exclusively on shipping to move its imports and exports. About 12,000 ships visit Australia each year, transporting nearly 350 tonnes of cargo.[10] Two thousand ships navigate the GBR annually, transporting everything from coal, nickel ores and alumina to raw sugar, sand, bauxite and oil.

Only 5 percent of these ships are tankers that carry crude or refined oil products. Yet tanker traffic is on the increase as the Queensland populations grow and demand increased refined oil products. Australia's future crude oil needs are expected to be supplied primarily from Indonesia and the Timor Gap.

Two routes of passage known as the *inner* and *outer* routes wind through the GBR's complex and dangerous maze of reefs, shoals, cays and islands, and connect with the Torres Strait at the region's northern end. The inner route follows the calmer waters between the reef and the Queensland coast. The outer route travels the deeper waters of the Coral Sea, seaward of the GBR, with passage into the lagoon between the reef and the coast at specified and limited narrow passages.

Eighty percent of all vessel voyages occur in the inner route, where other hazards include the 1,500 fishing vessels and 100,000 recreational and tourist vessels that congest shipping lanes in and around the ports. Whales and other large marine animals compound traffic problems.[11]

Given the natural dangers of navigation, the high volume of shipping traffic and the constrained routes, the risk of an oil spill in the GBR in the next 20 years is considered high.[12]

Australia combines national and international legislative muscle with national, state and territorial response and zoning plans to protect their natural wonder, supplemented by an aggressive educational and research program involving the public, major resource institutions, and private industry.

The entire GBR Marine Park is a marine protected area, zoned for multi-use with zones of no activity, restricted activity such as research, and limited commercial activity. Prohibited activities include oil drilling and production, mining, littering, spearfishing, and extracting large specimens of certain fish species. High impact activities require permits from the GBRMPA, which can be granted only after an extensive assessment is conducted.

Even with all these zoning controls in place, the largest marine protected area in the world faces a major problem: enforcement. The reef is so huge that catching violators is almost impossible. Even when violators are caught by the Australian Coast Guard, who are responsible for patrolling Marine Park waters, conviction is uncertain. For example, regulations prohibit fishing within 1,500 meters of the reef flat, but determining where the 1,500-meter line lies is often at question.[13]

PROTECTION UNDER INTERNATIONAL LAW

True to all navigation anywhere in the world, foreign ships enjoy the "right of innocent passage" under traditional law of the sea and, more specifically, under the Geneva Convention on Territorial Sea and Contiguous Zone, 1958. While Australia can protect its own seas to some extent, the real driving force behind control and protection of its GBR comes from Australia's ratification of international conventions, specific and specialized international status accorded the GBR, and adoption by Australia of necessary implementing legislation.

International law inhibits the right of innocent passage. The 1982 U.N. Convention on the Law of the Sea (UNCLOS) defines passage as "innocent" so long as it is not prejudicial to the peace, good order or security of the coastal state. As the coastal state, Australia can adopt laws relating to navigational safety, marine traffic control, preservation of the environment, and control and prevention of pollution from oil spills and other substances. Shipping lanes and shipping schemes can be prescribed, particularly where inherently dangerous or noxious substances are being shipped.

Australia has embraced the international system to ensure that its seas, territorial waters, and ports remain safe from pollution. Australia ratified and adopted significant International Maritime Organization conventions (see Chapters 17 and 18 for a full discussion of each of these international treaties), and then legislated to provide these protections for the GBR:

- *MARPOL 73/78, International Convention for the Prevention of Pollution from Ships:* Includes comprehensive control over pollution by ships at sea, including oil; criteria for safe operations, design and equipment controls, and reporting procedures; governs over 90 percent of world shipping tonnage. Given effect in Australia by the Protection of the Sea (Prevention from Ships) Act 1983 and the Navigation (Protection of the Sea) Amendment Act 1983.

- *CLC, International Convention on Civil Liability for Oil Pollution Damage:* Provides compensation from the shipowner for oil pollution damage, with insurance requirements to cover costs. Given effect in Australia by the Protection of the Sea (Civil Liability) Act 1981, with penalties of up to $100,000 for failure to carry appropriate insurance.

- *INTERVENTION, Intervention on the High Seas in Cases of Oil Pollution Casualties 1969* (and amending 1973 Protocol): Sets the right to take measures on high seas to protect coastline and related interests, including the well-being of living marine resources; can remove, destroy or take control of a ship that poses risk. Given effect in Australia by the Protection of the Sea (Powers of Intervention) Act 1981 with penalties of up to $50,000 for failure to comply with direction under the act and liability for all costs to government. Further protection under the Navigation Act 1912, with its right to remove a wreck when the threat of pollution is over.

- *OPRC, International Convention on Oil Pollution Preparedness, Response and Co-Operation:* Upgrades response capability of ships for oil spills, provides for emergency plans, reporting requirements, training and safety standards.

- *FUND, International Convention on the Establishment of an International Fund for Compensation for Oil Pollution Damage:* Provides additional compensation when CLC funds are insufficient, includes cleanup cost and economic losses.[14]

PSSA AND COMPULSORY PILOTAGE

In November 1990 the GBR was declared the first Particularly Sensitive Sea Area (PSSA) by the IMO. PSSAs are defined as areas "which need special protection through action by the IMO because of their significance for recognized ecological or socio-economic or scientific reasons and which may be vulnerable to damage by maritime activities."

The PSSA declaration places special protective measures on all shipping activities taking place in the GBR. Since 1991, loaded oil tankers, chemical tankers, liquefied gas carriers, and vessels 70 metres or more in length must carry a licensed pilot when taking passage through the inner route of the GBR between Cape York and Cairns, or when passing through the Hydrographers Passage, a much-used narrow passage between the inner and outer route.

The Australian Maritime Safety Act established the Australian Maritime Safety Authority (AMSA) in January 1991. This agency is charged to augment "the delivery of safety and other services to the Australian maritime industry." AMSA assumed five basic functions:

- Enhance maritime safety.
- Provide a national system of navigational services.
- Administer marine pollution prevention and response programs.
- Provide services to the maritime industry.
- Coordinate maritime search and rescue.

AMSA is responsible for administering applicable maritime international treaties, the Navigation Act 1912, and other maritime national legislation such as the Protection of the Sea (Prevention of Pollution from Ships) Act 1983. AMSA is the authority charged with enforcing compulsory pilotage, overseeing training and response to oil spills, and conducting Port State control and tanker surveillance inspections.

The master and owner of a vessel which navigates through a compulsory pilotage area without a pilot are both liable for felony prosecution and a maximum penalty following conviction of $50,000 for an individual and $200,000 for a corporation. The regulation applies whether the violator enters an Australian port on the same or a later trip, or returns at a later date with the same master. Since its inception, compliance with compulsory pilotage has risen from almost 90 percent to almost 100 percent, according to one study, although another study indicates that as much as 25 percent of foreign vessels traveling on the inner route still fail to meet the pilotage requirement. With the new upgraded pilot licensing system instituted in 1993 and administered by the AMSA, it is clear to the maritime community that compulsory pilotage through the GBR improves safety and lessens the possibility of oil spills.

The designation of the GBR as a PSSA allows the Australian government to take additional special preventive measures:

- Asserting certain discharge restrictions.
- Adopting routing measures, including prohibiting travel through certain areas.
- Introducing environmental fees, such as tolls for transit.
- Prohibiting certain activities such as offshore mineral exploration and fishing activities.

- Developing site-specific contingency plans to combat oil spills.
- Introducing Flag and Port State strict surveys of ships commensurate with increased risk to the environment due to age of vessels and other safety factors.
- Instituting vessel traffic management and reporting systems; for example, requiring vessels traveling through the reef to report every four hours.

The Australian government has implemented most of these measures. The international community assisted the GBR when, under MARPOL 73/78, they defined the designated area for no ship discharges of any type for the area between the Queensland coastline and 'nearest land' as a line drawn between coordinates on the outer edge of the GBR. This translates to a prohibition against discharge for oil tankers of up to 50 nautical miles from the outer edge of the GBR, and up to 200 nautical miles from the Queensland coast. For vessels other than oil tankers, the range of the prohibition extends between 12 and 162 nautical miles. Without this special definition, the usual distance from actual land in which discharges are prohibited is only 12 nautical miles.[15]

WORLD HERITAGE SITE/COMMUNICATION AND COMMITMENT

The GBR receives special protection through its inclusion on the UNESCO World Heritage List as a World Heritage Area, dating from October 1981. This listing requires the government of Australia to ensure the "protection, conservation and presentation of the Area and its transmission to future generations." Australia took this obligation seriously and developed a 25-year strategic plan for the GBR World Heritage Area, which includes the GBR Marine Park, certain waters outside the GBR Marine Park, and certain nearby islands.

The Strategic Plan is an example of conservation management at its best. Development involved all interested *stakeholders* (user groups, interest groups, state and local government agencies, 60 other organizations, and Aboriginal and Torres Strait islander groups), who met between August 1991 and plan adoption in April 1994. The plan is subject to review and revision every five years. The Australian government appropriates between A$60 million and A$100 million per year to implement the plan.

Underlying the entire plan is a basic goal: "An informed, involved, committed community," dedicated to maintaining a "healthy environment" for the GBR, not only for this generation, but for future Australians and the world. Education is paramount to achieving such an informed community.

GREAT BARRIER REEF
WORLD HERITAGE AREA STRATEGIC PLAN
THE 25 YEAR VISION

The illustration represents the vision of the future for the Area developed by participating organizations during the planning process. (Illustration courtesy of the Great Barrier Reef Marine Park Authority. First appeared in "The GBR World Heritage Area Stategy Plan." ©GBRMP.

One of GBRMPA's five departments is the Educational/Informational branch. This department produces brochures, publications and videos about ongoing research and reef projects, but its primary vehicle for communication is the aquarium at agency headquarters in Townsville, Australia. This aquarium operates at a deficit, but the loss is acceptable to the government. The aquarium contains one of the world's largest demonstration tanks of living coral and marine biodiversity, including a predator tank, and a tank where children of all ages can touch marine creatures such as sea cucumbers and sea stars. The facility provides weekend certification and other marine science courses, six-month in-depth courses, extension offices, summer camps for children, and a host of other school programs.[16]

OIL CONTINGENCY PLANNING

In 1973, the Australian government adopted a national strategy for responding to marine oil spills. This strategy was upgraded to its present format in 1993 as the National Plan to Combat Pollution of the Sea by Oil (National Plan), administered by AMSA. The National Plan maintains a national integrated government/industry organizational framework capable of responding effectively to oil pollution incidents, and to manage funding, equipment and training programs to support the plan. AMSA conducts a comprehensive training program. A shipping tax, the oil pollution levy of 4 cents per net registered ton per quarter for any commercial ship using Australian ports, provides funds for maintenance and administration. In 1991 the oil industry and the Australian Institute of Petroleum established the Australian Marine Oil Spill Centre in Geelong, where much of the oil spill response equipment is maintained.

The National Plan is supported by state and local contingency plans, such as the TORRESPLAN for the Torres Strait and the REEFPLAN for the GBR, administered by the Queensland government via the National Plan State Committee. Like the U.S. contingency plans, these plans prioritize oil spill response based on threats to human life, followed by habitat threat, then rare and endangered species, and finally threats to other natural resources based on environmental, social and economic factors and the specific spill incident. Unlike U.S. response, the government, not the owner/operator of the ship, responds to spills.[17]

Consider all the international support, the national legislation, educational activities on a statewide and territorial basis, compulsory pilotage, national and specialized response plans, criminal and monetary sanctions, all established to protect the GBR. Yet even with the most comprehensive system imaginable, Australians still do not rest easy that everything that can be done to prevent catastrophic oil discharge has been done. Without exception, those in charge at the GBRMPA responded to the oil spill question identically: *"We live in fear of an oil spill."*[18]

The importance of area contingency planning (ACP) cannot be emphasized enough. The ACP focuses on protection priorities and on prevention of oil spills. While diversion and control of shipping may be an answer for Australia's Great Barrier Reef and some other reef areas, one expert urges that the focus for other global communities living in coastal zones should be better ACP planning. Such planning, from the regional to the local level, is a focal point of OPA 90 and relevant international treaties: *"ACP response planning is the ultimate local community participation and stakeholder process."*[19]

THE POINT

Caroline Rogers, a devoted marine biologist, argues for action and erring on the side of more, not less protection:

"Managers will never have all the information they need. Decisions have to be made on the best available data, and the information must be objective. If any errors are made, they should be on the side of resource protection. Restrictions on visitation or use can always be relaxed if warranted. It is difficult to revive a dead reef. It is also far easier to start off with a good set of regulations rather than add them as an after thought." [20]

We can learn from those in the field, from scientists and managers who have devoted their lives to protecting the marine environment. The ultimate question is: "Do we have the political will to do what it takes to protect our coral reefs?"

WHAT *YOU* CAN DO

LOOK IN THE MIRROR

Surveys show that most of us claim to care about our environment. Many of us recycle, we do not litter, we take our used motor oil to an approved service station for disposal, we plant grass, trees and flowers, if we hunt, we do so only during designated seasons, and we contribute to charitable organizations devoted to saving land and marine wildlife and habitats. We like to believe these acts make us good citizens of our planet. On one level our actions are commendable. Our motives are correct. What we do not fully realize or appreciate is how significantly so many of our routine daily activities unknowingly contradict and undermine those good intentions by poisoning the sea or physically degrading water quality to dangerous levels for marine wildlife.

Rick Dawson of the USFWS summed up the problem of how to stop pollution from killing our oceans: "Look in the mirror. We are all part of the problem, and we don't even know it."[1] This statement conveys the basic truth that the *problem* of "good environmental stewardship" begins with each one of us, and the *solutions* begin with each of us, too.

The EPA has published a simple brochure with a powerful message. The pamphlet describes measures we as individuals can adopt in our everyday lives to halt toxins from pouring into our ocean from different land-based or *nonpoint* source pollution:

> *"Who is responsible? Nonpoint source pollution comes from a variety of land use activities, which means* we are all part of the problem! *The chemicals we use on our lawns, gardens, and crops; fluids leaking onto roadways from our cars; waste from our pets and farm animals; erosion from farm fields and construction sites, all contribute to nonpoint source pollution."*[2]

The EPA brochure offers some easy steps we can take to help make a difference:

- Reduce waste, reuse what you can, and recycle.
- Do not litter.

CHAPTER 23

- Use lawn and garden chemicals wisely, and apply the minimum you need.
- Use pump-out stations at marinas to dispose of human waste.
- Check your automobile for leaks and immediately repair those you find.
- Take used oil and fluids to a service station for disposal.
- Do not dump waste down storm drains.
- Collect and properly dispose of pet waste.[3]

We can undertake these and similar other tasks with a relatively insignificant increase in time, energy and awareness. Away from home, while vacationing elsewhere, we can also take care to protect reefs, mangroves and seagrass beds while we enjoy their many benefits. When visiting tropical or coastal areas:

- Practice safe boating. Do not anchor on reefs, in mangrove forests or in seagrass beds. Insist that your boat guide take care in anchoring near a reef.
- Use mooring buoys if they are present.
- Stay off the coral and keep your fins well away from reef creatures when diving or snorkeling.
- Limit the fish you spear or take from the reef, and respect local fishing regulations.
- Be careful when fueling your boat or changing engine oil to avoid needless spilling into the water.
- Do not buy coral, coral jewelry or "live rock" to take home as gifts or souvenirs. This only encourages the trade in these items. Leave the coral where it belongs – on the reef.

While all these suggestions are important, we live in a period in which the rapid deterioration of our coral reefs, mangrove forests and seagrass beds demands more from each of us.

Look in the mirror can also mean taking *political* action. We can choose to make political action a source of power and influence. *We* are the voters. *We* are the consumers of oil. *We* are the people our elected officials represent. The land and sea are ours to protect. Through our action, or inaction, *we* can determine whether oil will travel *first class*, rather than *business* or *economy class*. We do not all need to run for elected office to become involved. If we care and want to make a difference and influence policy, we can carry our commitment of personal environmentalism into the political arena, simply and effectively:

- Learn about environmental issues and support laws in your state that protect coastal areas, reefs, mangroves and seagrass beds.
- Write or fax clear, direct letters to your representatives in the U.S. House of Representatives and Senate, urging their support of upcoming proposed environmental legislation.
- Fill out and respond to well-drafted petitions addressed to appropriate legislators requesting them to vote against anti-environmental legislation coming before them when you receive notice from your environmental organization of choice. Sign and mail the card, letter or petition form.
- Pick up the telephone and call your elected representative in Washington, D.C. Telephone calls from constituents are logged by staff, and may take less time and expense to make than writing a letter.
- Make a point to attend local meetings with your legislator, and let him or her know what you think about issues.
- Participate in your hometown political arena.
- Discuss the issues with your friends and associates.
- Support environmental and conservation organizations of your choice that work to protect and preserve habitats, and generate public education and awareness about legislative issues.

A few specific suggestions can help you become a more active, informed participant in preserving our natural resource heritage:

- *Learn* more about environmental issues. (Learn about important wildlife habitats and support local efforts to protect critical areas.)
- *Become* part of the solution, and not part of the problem.
- *Support* legislation or organizations seeking solutions to these problems.
- *Report* illegal activities or resource violations.
- *Talk* to your friends and neighbors about what you have learned.[4]

In other words, become involved with projects in your local community and state. Safeguard basic environmental protection. Voice your concerns and opinions at meetings, hearings and fact-finding missions conducted by your local planning boards or agencies, and state advisory boards and commissions. Exercise your political muscle. You may be amazed, encouraged and empowered to discover the influence you can have at the grassroots level.

There is a larger sphere, the global arena. If we look in our mirror one more time, we can see what can be done to change the most basic patterns and thinking of our lives. Our consumption of fuel oil is increasing, not decreasing, annually. Most of us are not using alternative fuel sources. As one architect says, "We are still 'on the grid.'"

Sir Shridath Ramphal, president of the World Conservation Union (IUCN), made an eloquent appeal for action at the highest level with a call to our *will to lead*:

> *"But do we yet have the political will to implement the solutions that are required? Are consumers in the North willing to curb their overuse of energy, and emission of greenhouse gases that already over-tax resilience of Nature? Are they willing to alter the terms of trade, and reverse the economic flows, so that development in the South is stimulated rather than continually undermined? Are governments everywhere willing to see that unless local communities are empowered to plan and manage their own resources, inequity, exploitation and a waste of natural resources is inevitable? And are the leaders of the conservation movement willing to put human needs at the forefront of their concerns, and insist that while development must be conservation-based, conservation must also be people-centered?"[5]*

Because of OPA 90, COFRs, and the vast array of regulation enforced by the USCG, the U.S. is a world leader in support of oil-free oceans. It is vitally important that U.S. citizens stand firm on this strong legislation and oppose efforts to weaken it or water it down. Demand that the *polluter pays*, and request that the international arena follow the U.S. lead in insisting on safe oil transport.

REEFS OF THE FUTURE

If our waters continue to become increasingly polluted in the absence of controls, if coastal populations continue to over-use and abuse their reef resources in the absence of conservation awareness and education, if disastrous spills and gross contamination persist in striking the Caribbean and other shallow seas, natural coral reefs may disappear from our oceans altogether. Earth may one day become a world of artificial reefs.

There are programs that help injured reef corals regenerate. Hector Guzman, a biologist with the Smithsonian Tropical Research Institute in

Artificial reef module. (Reprinted by permission of Dept. of Natural and Environmental Resources, San Juan, Puerto Rico.)

Panama, has transplanted coral to damaged reefs around Colombia, Costa Rica and Panama, using thousands of stakes wired with species of stony coral. After 18 months, the experiments produced colonies large enough to survive. But Guzman's painstaking work and other efforts will not succeed if man-made and natural sources of reef degradation continue at present rates. Guzman's results depend on a delicate, precise set of conditions, including clean, clear waters free of sediment and pollutants with suitable surfaces for corals to attach themselves.[6]

Another type of "reef" is being deployed around the world, a reef made from cement and steel. Puerto Rico is experimenting with this new artificial reef. The fisheries division of the DNR, under the leadership of Jose Berrios, has conducted a successful experiment to increase fish habitats by using artificial reef structures off the coast of mainland Puerto Rico in Humacao and Aguadilla. The artificial reefs, called *modules,* are composed of grey concrete scrap pipes bound together by packing metal strap. The cost of this unsightly storm pipe assemblage is minimal.

The Humacao modules were deployed near an old pier, far removed from a natural reef, where few fish were present. The Aguadilla cement reef was dropped near another pier in about six feet of water. Both the sites have been monitored for diversity and abundance of fish since April 1994. The modules appear to be working, at least for fishery purposes. Where formerly only small juveniles of some limited species existed, there are now multi-sized fish of many species.[7]

Artificial reefs are also used on the site of vessel groundings, such as the *Berman*, where the barge caused almost total reef destruction in a swath 150 meters long by 60 meters wide, and partial reef destruction from rubble thrown about in the midsection of the large grounding site. In such a situation where the reef has suffered significant and irreversible long-term impacts, the choices are limited. Ideally, restoration activities such as Guzman's transplanting process would be undertaken directly over the affected zones. Where climatic and oceanographical factors make this impractical in terms of cost/benefits, durability of the deployed structures and safety, the use of an artificial reef some distance from the grounding site is a secondary option. The purpose of the artificial reef in a natural resources damage situation is to restore the lost services supplied by the reef assemblage prior to the physical disruption of the site.[8]

Unfortunately, few if any reef corals are actually growing on these structures, and those are limited to several types, such as gorgonians and sea fans. To date these concrete structures do not appear to encourage the creation and colonization of a live stony coral reef with its wealth of diversity and sea creatures.[9] Although these assemblages have been used throughout the world for the past two decades, scientists warn about *artificial habitat ecology*, its proper use in fisheries work, mitigation of environmental damage, effectiveness, and value under different circumstances. Too little is known about this young science to make meaningful generalizations.[10]

While we can applaud fishery projects for stimulating fish populations and generating food sources and recreation for people, the cement

Reef of the future? (Reprinted by permission of Dept. of Natural and Environmental Resources, San Juan, Puerto Rico.)

Coral structure. (Reprinted by permission of World Wildlife Fund.)

artificial reef is no replacement for the miracle of a living natural coral superstructure. Photographs of artificial reefs provide a stark contrast to the spectacular natural reefs many local residents and visitors worldwide treasure. The modules serve a purpose when other means have failed, but surely no one would prefer an artificial reef to the real thing.

We are in a race with time to save our natural coral reefs. People have caused the death of 5 percent to 10 percent of the world's coral reefs already. Scientists predict within the next 40 years we may eliminate another 60 percent of the coral reefs now in existence. The latest research suggests that reefs may be disappearing at an even faster rate, with a projected loss of 30 percent in the next 20 years.[11]

Legislation that passes today becomes law that stays on the books for decades. The U.S. has led the shipping industry by requiring compliance with stringent regulations that help prevent accidental discharges of disastrous quantities of oil into American and international waters. If we as a country do not stay the course, we may well be faced with exactly what the supporters of OPA 90 sought to avoid: "I do not want some future headline to read, 'Washington Studies While the Oil Spills.'"[12]

You and I are in charge of our future. We can prevent those who represent us from taking detrimental legislative action that may curtail OPA 90's *polluter pays* policy. Our actions and inaction each day in our homes and backyards, in our local communities and larger political systems, and in the global arena, may well determine the ultimate survival or demise of the vital coral beings that have inhabited our planet for the past 600 million years.

ENDNOTES

ENDNOTES

CHAPTER 1 - ANATOMY OF AN OIL SPILL: HOW DID *BERMAN* HAPPEN?

1. Doyle, Capt., "Casualty Investigation Report, Grounding of the Tank Barge *Morris J. Berman*," USCG, San Juan, Puerto Rico, 5/16/94.

2. Department of Environmental Conservation, "Decision and Order," *State of New York v. Berman Enterprises, et.al.*, R2-3291-90-11, New York, 3/25/91.

3. Department of Environmental Conservation, "Decision and Order," 3/25/91.

4. Department of Environmental Conservation, "Hearing Report," *New York State v. Berman Enterprises, et.al.*, R2-3291-90-11, 1991.

5. USCG, "Casualty Investigation Into the Circumstances Surrounding the Grounding at Punta Escambrón, Puerto Rico, January 7, 1994," San Juan, Puerto Rico, 1/12/94, p. 56.

6. U.S. Coast Guard, "Towing Vessel Inspection Study," Merchant Vessel Inspection and Documentation Division, 1994, p. 5; Dept. of Environmental Conservation, "Hearing Report," 1991.

7. USCG, "Marine Safety Information System Reports," Systems Support Branch, Washington, D.C., 8/31/81 - 1/15/94.

8. Barker, LCDR, USCG, Interview, 10/5/94; Robert Ross, Capt., USCG, Interview, 2/15/95; Edwin Stanton, CDR, USCG, Interview, 9/27/94.

9. Tobin, William A., "Metallurgical Report of the FBI Laboratory, Escambrón Beach Oil Spill," U.S. Dept. of Justice, Washington, D.C., 1/31/94.

10. USCG, "Casualty Investigation Hearing Transcript," 1/11/94, p. 156.

11. Kehoe, Daniel, Capt., "Answers to Interrogatories in USCG Casualty Investigation," 1/29/94.

12. USCG, "Casualty Investigation Hearing Transcript," 1/12/94, p. 60.

13. USCG, "Casualty Investigation Hearing Transcript," 1/12/94, p. 62.

14. USCG, "Casualty Investigation Hearing Transcript," 1/12/94, p. 65.

15. Doyle, Capt., "Casualty Investigation Report, Grounding of the Tank Barge *Morris J. Berman*," USCG, San Juan, Puerto Rico, 5/16/94.

16. USCG, "Casualty Investigation Hearing Transcript," 1/12/94, p. 70.

17. USCG, "Casualty Investigation Hearing Transcript," 1/12/94, p. 72; Doyle, Capt., USCG, "Casualty Investigation Report," 5/16/94, p. 12.

18. USCG, "Casualty Investigation Hearing Transcript," 1/13/94, p. 108.

19. Doyle, Capt., "USCG Casualty Investigation Report," 5/16/94, pp. 12-13.

20. USCG, "Casualty Investigation Hearing Transcript," 1/12/94, p. 21.

21. USCG, "Casualty Investigation Hearing Transcript," 1/12/94, p. 138.

22. USCG, "Casualty Investigation Hearing Transcript," 1/12/94, pp. 31-32.

23. USCG, "Casualty Investigation Hearing Transcript," 1/12/94, p. 137.

24. Doyle, Capt., "USCG Casualty Investigation Report," 5/16/94, pp. 18-19; Tobin, "Metallurgical Report of the FBI Laboratory," 1/31/94.

25. USCG, "Casualty Investigation Hearing Transcript, 1/12/94, pp. 137, 139.

26. Tobin, "Metallurgical Report of the FBI Laboratory," 1/31/94, p. 9.
27. Robert Ross, Capt., USCG, reviewer comments, 8/12/96.
28. USCG, "Casualty Investigation Hearing Transcript," 1/12/94, p. 125.
29. USCG, "Casualty Investigation Hearing Transcript," 1/13/94, p. 125.
30. USCG, "Casualty Investigation Hearing Transcript," 1/13/94, pp. 85, 87.
31. USCG, "Casualty Investigation Hearing Transcript," 1/12/94, p. 148.
32. USCG, "Casualty Investigation Hearing Transcript," 1/11/94, p. 106.
33. Doyle, Capt., "USCG Casualty Investigation Report," 5/16/94, pp. 4, 5.
34. USCG, "Casualty Investigation Hearing Transcript, 1/11/94, pp. 66-67.
35. USCG, "Casualty Investigation Hearing Transcript," 1/11/94, pp. 39, 71.
36. Doyle, Capt., "Casualty Investigation Hearing Report," 5/16/94, p. 6
37. USCG, "Towing Vessel Inspection Study, Final Report," Washington, D.C., 1994; *Oil Spill Intelligence Reporter*, "USCG Issues Final Rule on Operational Measures for Single-Hull Tankers," Vol. XIX, No. 30, August 8, 1996.
38. Robert Ross, Capt., Interview, 1/11/95.
39. Lee Crockett, Interview, 9/5/95; U.S. Congressional Record - House, "Towing Vessel Safety Act," March 16, 1994; Chafee, John Sen., S. 1730, May 7, 1996; OSIR, August 8, 1996.
40. Ross, Robert, Capt., "The Response to the T/B *Morris J. Berman* Major Oil Spill," San Juan, Puerto Rico, 8/25/95, Enclosure 3.

CHAPTER 2 - THE WAR ON SPILLED OIL

1. Mark Miller, Interviews, 8/25/96, 8/26/96.
2. Robert Ross, Capt., Interviews, 1/11/95, 2/1/95, and reviewer comments, 8/96.
3. Ross, Robert, Capt., "The Response to the Barge T/B *Morris J. Berman* Spill," San Juan, Puerto Rico, 8/25/95.
4. Edwin Stanton, CDR, Interview, 5/31/94.
5. Robert Ross, Capt., Interview, 1/11/95.
6. Aileen Velazco-Dominguez, Interview, 9/21/94.
7. Ross, Robert, Capt, "The Response" 8/25/95; Robert Ross, Capt., reviewer comments, 8/96; Dan Hamson, Interview, 7/23/94; Garamone, Matthew, "A comprehensive approach to incident management," *Hazmat World*, March, 1994: 70-73.
8. Wilkinson, Clive R. and Buddemeier, R.W., "Global Climate Change and Coral Reefs: Implications for People and Reefs," UNEP-IOC-ASPEI-IUCN Global Task Team, 1994.
9. Ross, Robert, Capt., "The Response, " 8/25/95.
10. Ross, Robert, Capt., "The Response," 8/25/95; Robert Ross, Capt., Interview, 1/11/95.
11. Robert Ross, Capt., reviewer comments, 8/96.
12. White, Dr. I. C., "Major Oil Spills - There is No Miracle Cure," ITOPF, n.d: 16-19.

13. USCG Office of Marine Safety, Security and Environmental Protection, "An Assessment of Towing Vessel Manning Requirements," July 1994.

CHAPTER 3 - OFFSHORE RESPONSE AND SINKING THE *BERMAN*
1. Kevin Peters, Interview, 10/10/94.
2. Edwin Stanton, CDR, Interview, 5/31/94.
3. Kevin Peters, Interview, 10/10/94.
4. Ross, Robert, Capt., "The Response to the T/B *Morris J. Berman* Oil Spill, " USCG, San Juan, Puerto Rico, 8/25/95.
5. Paul Hankins, Interview, 10/11/94; Mark Miller, Interview, 8/25/96; U.S. Navy, "Final Report *Morris J. Berman* Spill, San Juan, Puerto Rico," Naval Sea Systems Command, Arlington, Virginia, 7/26/94.
6. O'Brien, James L. and J. Gallagher, "The Mother of All Oil Spills and the Dawhat Ad Dafi," reported in proceedings of the 1993 International Oil Spill Conference, 1993.
7. Richard Hooper, Interview, 10/14/94.
8. NSFCC, "USCG National Strike Force 1992 Year in Review," USCG, Elizabeth City, N.C., 1992.
9. Jerry Hubbard, Lt. Jg., Interview, 8/2/94.
10. Edwin Stanton, CDR, Interview, 8/96; Paul Hankins, Interview, 8/96.
11. Paul Hankins, Interview, 10/11/94.
12. Ross, Robert, Capt., "The Response" 8/25/95, Oil Budget Figure 6.
13. Jerry Hubbard, Lt. Jg., Interview, 8/2/94.
14. Jerry Hubbard, Lt. Jg., Interview, 8/2/94.
15. Mark Miller, Interview, 8/25/94.
16. National Response Corporation, "Logistics Summary - Oil Spill Response - San Juan, Puerto Rico, New York, 1/28/94.
17. Mark Miller, Interview, 8/26/94.
18. USCG, "News Release, Release #17," San Juan, Puerto Rico, 1/15/94.
19. Ross, Robert, Capt., "Natural Resources Report - Barge *Morris J. Berman* Spill," USCG, San Juan, Puerto Rico, 2/3/94.
20. Robert Ross, Capt., Interview, 2/1/95.
21. A. Carlo Calanni, Interview, 7/96.
22. Ross, Robert, Capt., "Natural Resources Report," 2/3/94; U.S. Navy, "Final Report, *Morris J. Berman* Spill," Naval Sea Systems Command, San Juan, Puerto Rico, 7/26/94.
23. Richard Hooper, Interview, 10/14/94.
24. Robert Ross, Capt., Interview, 1/11/95, Edwin Stanton, CDR, Interview, 9/27/94.
25. Robert Ross, Capt., Comments, 8/96.
26. Hankins, Interview, 10/11/94, U.S. Navy, "Final Report," Naval Sea Systems Command, 7/26/94.
27. Ross, Robert, Capt., "The Response," 8/25/95.

28. Nichols, J.A., Draft Critique, ITOPF, London, U.K., 8/17/94.
29. Robert Ross, Capt., Interview, 1/11/95; Robert Ross, Capt., Reviewer comments, 8/96.
30. Robert Ross, Captain, Interview, 2/1/95.
31. Robert Ross, Capt., "The Response," 8/25/95.
32. Robert Ross, Capt., "The Response," Oil Budget, Figure 6, 8/25/95. Robert Ross, Capt., reviewer comments, 8/96.
33. Robert Ross, Capt., Interview, 2/1/95.

CHAPTER 4 - RESCUING THE LIVING KINGDOM
1. Mignucci-Giannoni, A.A.M.V. Pauley and D.P. Moore, "Preliminary mortality assessment, rescue and rehabilitation of all wildlife affected by the barge *Morris J. Berman* spill in SJ, PR. La Parguera, PR," unpublished final contract report of Red Caribeña de Varamientos (Caribbean Stranding Network) to the Dept. of Natural and Environmental Resources, Commonwealth of Puerto Rico, 7/15/94, 29 pp.
2. Mignucci-Giannoni, et. al., 7/15/94, p. 16.
3. James Timber, Interview, 6/22/94.
4. James Timber, Interview, 9/21/94.
5. Vicente, Vance P., "Oil Spill Report/Cumulative Information," SEFSC/WMFS/NOAA, 1/7 - 1/20/94, San Juan, Puerto Rico; Vicente, Vance P., "Field Notes," SEFSC/NOAA, 1/20/94.
6. Bradford Benggio, LCDR, Interview, 7/7/94.
7. Felix Lopez, Interview, 6/27/94.
8. Vicente, Vance P., "Field Notes," SEFSC/NOAA, San Juan, Puerto Rico. 1/16-1/20/94; Dan Hamson, Interview, 7/23/94.
9. Felix Lopez, Interview, 6/27/94.
10. Antonio Mignucci-Giannoni, Interview, 6/94.
11. Mignucci, et.al., 7/15/94.
12. Rick Dawson, Interview, 10/94.
13. Rick Dawson, Interview, 10/94.
14. Ross, Robert, Capt., "Natural Resources Report, Barge *Morris J. Berman* Spill," USCG, San Juan, Puerto Rico, 2/3/94.
15. Jorge Saliva, Interview, 7/11/94.
16. Jorge Saliva, Interview, 7/11/94.
17. Antonio Mignucci-Giannoni, Interview, 6/94.

CHAPTER 5 - SHORELINE CLEANUP PART I:
BEACHFRONT, SEA BOTTOM AND ROCKY COASTLINE
1. Benggio, Brad, "Draft Scientific Support Coordinator's Report - *Berman* Spill," NOAA, 1/7/94.
2. Robert Ross, Capt., Interview, 1/11/95.
3. Benggio, Brad, "Draft Scientific Support Coordinator's Report - *Berman*

Spill," NOAA, 1/7/94.

4. Pablo Roque, Lt., Interview, 7/18/94.

5. Felix Lopez, Interview, 6/27/94.

6. Benggio, Brad, "Draft SSC Incident Report," 1/7/94; Henry, Charles B., "Institute for Environmental Studies Chemistry Report: IES/RCA T94-10," Louisiana State University, Baton Rouge, LA, 1/31/94.

7. Ross, Robert, Capt., "Submerged Oil Recovery Operations Tank Barge *Morris J. Berman* Spill," USCG, San Juan, Puerto Rico, 7/26/94.

8. Ross, Robert, Capt., "Memo, Pollution and Response for Low API Gravity Oils, USCG, San Juan, Puerto Rico, 4/14/94.

9. Ross, Robert, Capt., "Natural Resources Report," 2/3/94.

10. Brad Benggio, LCDR, Interview, 7/7/94.

11. Robert Ross, Capt., reviewer comments, 8/96.

12. Felix Lopez, Interview, 6/27/94.

13. Nelson Pereira, Petty Officer, Interview, 7/19/94.

14. Bradford Benggio, LCDR, Interview, 7/7/94.

15. Robert Ross, Capt., Interview, 8/96.

16. Robert Ross, Capt., Interview, 1/11/95.

17. Vance Vicente, Interview, 6/30/94, Brad Benggio, Interview, 7/7/94, Ross, Robert, Capt., "Natural Resources Report," 2/3/94.

18. Ross, Robert, Capt., "Natural Resources Report," 2/3/94.

19. Edwin Stanton, CDR, Interview, 9/27/94.

20. Mayra Garcia, Interview, 9/21/94.

21. Aileen Velazco-Dominguez, Interview, 9/21/94.

22. Aileen Velazco-Dominguez, Interview, 9/21/94.

23. Freddie Aledo, Interview, 9/20/94.

24. Nelson Pereira, Petty Officer, Interview, 7/19/94.

25. Brad Benggio, LCDR, Interview, 7/7/94.

26. Robert Ross, Capt., Interview, 1/11/95.

27. Brad Benggio, LCDR., Interview 7/7/94, Felix Lopez, Interview, 6/27/94.

28. Ross, Robert, Capt., "Natural Resources Report," 2/3/94.

29. Susan Soltero, Interview, 6/1/94.

30. Ross, Robert, Capt., "Submerged Oil Recovery Operations," 7/26/94.

31. USCG, "*Morris J. Berman* Oil Spill," News Release #69-94, 11/4/94; Ross, Robert Capt., "Submerged Oil Recovery Operations," 7/26/94.

32. Benggio, Brad, "Staff Briefing Report- *Morris J. Berman* Oil Spill," NOAA Scientific Support Coordinator, 2/18/94; Ross, Robert, Capt., "Submerged Oil Recovery Operations," 7/26/94.

33. Benggio, Brad, "Staff Briefing Report," 2/18/94; Ross, Robert, Capt., "Submerged Oil Recovery," 7/26/94.

CHAPTER 6 - SHORELINE CLEANUP PART II: SAND, SLUDGE AND SPECIAL PROBLEMS

1. Ross, Robert, Capt., "Submerged Oil Recovery Operations - Tank Barge *Morris J. Berman* Spill," USCG, San Juan, Puerto Rico, 7/26/94; Edwin Stanton, CDR, Interview, 9/27/94; Benggio, Brad, "Draft SSC Incident Report," NOAA, 1/7/94; Ross, Robert, Capt., "Natural Resources Report, Barge *Morris J. Berman* Spill," USCG, San Juan, Puerto Rico, 2/3/94.

2. Rick Good, Interview, 7/27/94; Ross, Robert, Capt., "The Response to the T/B *Morris J. Berman* Major Oil Spill, USCG, San Juan, Puerto Rico, 8/25/95.

3. Genaro Torres, Interview, 9/28/94; Ross, Robert, Capt., "The Response," 8/25/95.

4. Edwin Stanton, CDR., Interview, 5/31/94.

5. Ross, Robert, Capt., "The Response," 8/25/95.

6. Rick Good, Interview, 7/27/94.

7. Rick Good, Interview, 7/27/94; Lourdes Ferra, Interview, 9/21/94.

8. Ross, Robert, Capt., "Heritage Resources Report II, 1/27 - 4/94," USCG, San Juan, Puerto Rico.

9. McKinley, Audrey, LCDR. and Agamemnon Gus Pantel, "Managing Heritage Resource Protection: A Case Study from the *Berman* Spill Response," USCG, Paper ID #47, reported in proceedings of 1995 International Oil Spill Conference, 2/95.

10. Agamemnon Gus Pantel, Interviews, 8/2/94, 9/20/94.

11. National Park Service, "Impact on Cultural Resources, S.J., P.R., U.S. Dept. of the Interior, 7/94; Ross, Capt., "Heritage Resources Report, Vol. II," 1/27-4/94.

12. McKinley and Pantel, "Managing Heritage Resources, 2/95.

13. Dan Hamson, Interview, 7/23/94; McKinley and Pantel, "Managing Heritage Resources", 2/95.

14. Agamemnon Gus Pantel, Interview, 9/20/94.

15. Agamemnon Gus Pantel, Interview, 9/20/94.

16. Agamemnon Gus Pantel, Interview, 9/20/94.

17. McKinley and Pantel, "Managing Heritage Resources," 2/95.

CHAPTER 7 - VIEW FROM THE TOP: MANAGING AN OIL SPILL

1. USCG, "News Release - Oil Spill Cleanup Update #11 - 2 P.M.," San Juan, Puerto Rico, 1/12/94.

2. Puerto Rico and U.S. Virgin Islands Oil Contingency Plan, n.d.

3. Matthew Garamone, Interview, 8/30/94; EPA, "Overview of the Emergency Response Program, Publication 9360.0-25, Washington, D.C., April, 1992; USCG, the Coastal Area Contingency Plan for Puerto Rico and U.S. Virgin Islands, 1994, San Juan, Puerto Rico, 1994.

4. Aldaberto Bosque, Interview, 5/31/94; Puerto Rico and U.S. Virgin Island Area Oil Spill Contingency Plan, n.d.

5. OPA 90.
6. Robert Ross, Capt., Interview, 2/1/95.
7. Ross, Robert, Capt., "The Response to the T/B *Morris J. Berman* Oil Spill," San Juan, Puerto Rico, 8/25/95; Ross, Robert, Capt., reviewer comments, 8/96.
8. Edwin Stanton, CDR, Interview, 3/1/95; Robert Ross, Capt., Interviews, 1/11/95, 2/11/95.
9. *Golob's Oil Pollution Bulletin*, "Jury convicts parties involved in *Morris J. Berman* incident," Vol. VIII, No. 9, 5/3/96.
10. Felix Lopez, Interview, 6/27/94.
11. Robert Ross, Capt., Interview, 1/11/95; Ross, Robert, Capt., "The Response," 8/25/95.
12. Ross, Robert, Capt., "The Response," 8/25/95.
13. Robert Ross,Capt., Interview, 1/11/95.
14. Robert Ross, Capt., Interviews, 1/10/95, 1/11/95, 2/1/95.
15. Ross, Robert, Capt., "The Response," 8/25/95.

CHAPTER 8 - THE CORAL REEF QUESTION

1. White, I. C. "Environmental Damage in the Context of the International Compensation Regimes," presented at ITOPF Seminar on NOAA Proposed NRDA Regulations, ITOPF, 1994.
2. Jerry Barnett, Interviews 6/1/94 and 7/20/94; International Maritime Organization, "Oil pollution declines as shipping measures take effect," *IMO News*, No.3:1993.

CHAPTER 9 - THE CORAL REEF AND ITS INHABITANTS

1. *National Audubon Society Field Guide to North American Seashore Creatures*: New York, Alfred A. Knopf, 1994, p. 24; U.S. Department of State, "The International Coral Reef Initiative Working Paper," Washington, D.C., October 20, 1995.
2. Wilkinson, Clive R. & Robert W. Buddemeier, "Global Climate Change and Coral Reefs: Implications for People and Reefs," UNEP-IOC-ASPEI-IUCN, 1994, p. 28.
3. Gilberto Cintron-Molero, Interviews 7/12/94, 9/13/94.
4. Peter Hulm and John Pernetta, "Reefs at Risk - Coral reefs, human use and global climate change," Marine and Coastal Programme of IUCN-The World Conservation Union, Gland, Switzerland, October, 1993: 24 pp; Wilkinson and Buddemeier, p. 2.
5. Hulm and Pernetta, p. 7.
6. Brown, Barbara E. and John C. Ogden, "Coral Bleaching," *Scientific American*, January, 1993: p. 65.
7. Brown and Ogden, p. 65; Hulm and Pernetta, p. 8; Caroline Rogers, Interview, 1/4/95.

8. Orr, Katherine, (ed.) Leslie Y. Lin, "Teacher's Guide to Coral Reef Teaching Kit," Washington, D.C., World Wildlife Fund, 1986, p. 22.

9. Littler, Diane Scullion, Mark M. Littler, Katina E. Bucher and James N. Norris, *Marine Plants of the Caribbean*, Washington, D.C., Smithsonian Institute Press, 1989, pp. 19, 133.

10. Wells, Sue and Nick Hanna, *The Greenpeace Book of Coral Reefs*, New York, Sterling Publishing Co., Inc., 1992, p. 18.

11. Stafford-Deitsch, Jeremy, *Reef: A Safari Through the Coral World*, San Francisco, California, Sierra Club Books, 1993, p. 166.

12. Kaplan, p. 112; Sylvia A. Johnson, *Coral Reefs*, Minneapolis, Lerner Publications Co, 1984, pp. 9, 10.

13. Kaplan, p. 113.

14. Orr, Katherine, p. 22.

15. Wells and Hanna, p. 21.

16. Kaplan, p. 113.

17. Orr, Katherine, p. 22.

18. Eugene A. Shinn, Interview, 6/24/94; Shinn, Eugene A., "Geology and Human Activity in the Florida Keys," U.S. Geological Survey, Energy and Marine Ecology, October, 1993.

19. Bruckner, Andrew, "The Threatened Coral Reefs of Jamaica: The Problem and Potential for Conservation," *Jamaica Naturalist*, n.d.

20. Stoddard, D.R., "Ecology and morphology of recent coral reefs," *Biol. Rev.* 44(4):433-498, 1969.

21. Wells and Hanna, p. 18.

22. Goenega, Carlos, *The State of Puerto Rican Corals: An Aid to Managers*, Mayaguez, Puerto Rico, University of Puerto Rico, May 1991, p. 10.

23. Gilberto Cintron-Molero, Interview, 7/12/94; Wells and Hanna, pp. 24-25.

24. Wells and Hanna, p. 26.

25. Aileen Velazco-Dominguez, Interview, 9/21/94; LaPointe, Brian E., "Caribbean Coral Reefs: Are They Becoming Algal Reefs?," *Sea Frontiers* 35:2, March-April 1989, 84-85; Kaplan, Eugene H., *A Field Guide to Coral Reefs - Caribbean and Florida*, Boston, Peterson Field Guide Series, 1982, pp. 101-103; Gilberto Cintron-Molero, reviewer comments, 8/20/96.

26. Allen, William H., "Increased Dangers to Caribbean Marine Ecosystems," *BioScience* 42:5, May 1992, p. 332; Wells and Hanna, pp. 10, 11; Stafford-Deitsch, p. 47.

27. Wells and Hanna, p. 23; Cintron-Molero, Gilberto, *Restoring the Nation's Marine Environment*: (ed.)by Gordon W. Thayer, Maryland Sea Grant College, 1992, pp. 227-228; Gilberto Cintron-Molero, reviewer comments, 8/20/96.

28. Wells and Hanna, P. 24; Australian Dept. of Transportation, "Great Barrier Reef and Torres Straight Shipping Study, Vol. 2," CRC Reef Research Center, December, 1994.

29. Susan Drake, Interview, 10/19/94.

CHAPTER 10 - NATURAL THREATS TO CORAL REEFS

1. Aileen Velazco-Dominguez, Interview, 9/21/94.
2. Eugene Shinn, Interview, 6/24/94; Shinn, Eugene A., "Coral Reef Recovery In Florida and the Persian Gulf," *Environmental Geology* 1 (1976): 245; Shinn, Eugene A., "What is Really Killing the Corals," *Sea Frontiers* Vol. 35, No. 2 (March-April 1989): 73-75.
3. Bruckner, Andrew, "The Threatened Coral Reefs of Jamaica: The Problem and Potential for Conservation," *Jamaican Naturalist* (n.d.), p. 19; Rogers, Caroline S., Larry N. McLain, Craig R. Tobias, "Effects of Hurricane Hugo (1989) on a Coral Reef in St. John, USVI," Marine Ecol. Prog. Ser. 78 (1991): 189-199; Rogers, Caroline S., "A Matter of Scale: Damage From Hurricane Hugo (1989) to U.S. Virgin Islands Reefs at the Colony, Community, and Whole Reef Level," Proceedings of the Seventh International Coral Reef Symposium, Guam, 1992, Vol. 1: 127-133; Caroline Rogers, Interview, 9/23/94.
4. Allen, William H., "Increased Damages to Caribbean Marine Ecosystems," *BioScience*, Vol. 42, No. 5 (May 1992): 330-335.
5. LaPointe, Brian E., "Caribbean coral reefs: Are they becoming algal reefs?" *Sea Frontiers*, Vol. 35, No. 2 (March-April, 1989): 82-91; Hughes, Terence P., "Catastrophes, Phase Shifts, and Large-Scale Degradation of a Caribbean Coral Reef," *Science*, Vol. 265 (Sept. 9, 1994): 1547-1551.
6. Shinn, (March-April 1989,) pp. 78, 79, 81; LaPointe, pp. 87, 89, 91; Bruckner, pp. 20, 22.
7. Wells, Sue and Nick Hanna, *"The Greenpeace Book of Coral Reefs,"* New York, Sterling Publishing Co., Inc., 1992, pp. 61-63.
8. Aileen Velazco-Dominguez, Interview, 9/21/94; Smith, S. V. and R. W. Buddemeier, "Global Change and Coral Reef Ecosystems," Annu. Rev. Ecol. Syst. 1992. 23:89-118; Wells and Hanna, p. 56; Hulm, Peter and John Pernetta, "Reefs at Risk: Coral Reefs, Human Use and Global Climate Change," IUCN: Oct. 1993, pp. 17-19; Wilkinson and Buddemeier, pp. 12-15.
9. Porter, James W., W.K. Fitt, H.J. Spero, C.S. Rogers, and M.W. White, "Bleaching in reef corals: Physiological and stable isotopic responses," Proc. at Acad. Sci. USA Vol. 86, pp. 9342-9346, December 1989, p.9345.
10. Hulm and Pernetta, pp. 17-18; Wilkinson and Buddemeier, pp. 14 and 15; Smith and Buddemeier, pp. 102-103.
11. Porter, pp. 9342-9346: Brown, Barbara E. and John C. Ogden, "Coral Bleaching," *Scientific American*, Jan. 1993: 64-70; Smith and Buddemeier, p. 94; Hulm and Pernetta, 17-19.
12. Hulm and Pernetta, p. 15.

CHAPTER 11 - MAN-MADE THREATS TO CORAL REEFS

1. Hulm, Peter and John Pernetta, "Reefs at Risk: Coral Reefs, Human Use and Global Climate Change," IUCN, Oct. 1993, p.12.

2. GESAMP Reports and Studies No. 50, "Impacts of Oil and Related Chemicals and Waste on the Marine Environment," Joint Group of Experts on the Scientific Aspect of Marine Pollution, London, International Maritime Organization, 1993, p. iii.

3. EPA, "The Facts: San Juan Bay Estuary System, Puerto Rico," The National Estuary Program, n.d.

4. Smith, S.V. and R.W. Buddemeier, "Global Climate Change and Coral Reef Ecosystems," Annu. Rev. Ecol. Syst., 1992, 231:89-118; Gilberto Cintron-Molero, reviewer comments, 8/20/96.

5. Wilkinson, Clive R. and Robert W. Buddemeier, "Global Climate Change and Coral Reefs: Implications for People and Reefs," UNEP-IOC-ASPEI-IUCN Global Task Team, 1994, p. 19.

6. Hughes, Terrence P., "Catastrophes, Phase-Shifts, and Larger-Scale Degradation of a Caribbean Coral Reef," *Science*, Vol. 265, Sept. 9, 1994: 1547-1551.

7. Montalbano, "U.N.: Oceans Over Fished at Alarming Rate," *The Denver Post*, March 26, 1995.

8. *Traffic USA*, "Coral Commerce Concerns Conservationists," World Wildlife Fund, Vol. 15, No. 1, January, 1996: 4 pp.

9. Allen, William H., "Increased Damages to Caribbean Marine Ecosystems," *Bioscience* Vol 42, No. 5, May, 1992: 330-335.

10. Rogers, Caroline S., "Efforts to Balance Marine-based Tourism with Protection of Coral Reefs and Seagrass Beds in a National Park," in Needham, B. (ed.), Publ. No. 1001, Coastal Resources Center, University of Rhode Island, 1991, pp. 71-82.

11. *St. Thomas This Week*, Vol. XXXV, No. 53, December 31, 1994.

12. EPA, n.d.

13. Australian Department of Transportation, "Great Barrier Reef and Torres Strait Shipping Study, Vol. II," Cooperative Research Centre (CRC) Reef Research Centre, December 1994, pp. 4-8.

14. Wells, Sue and Nick Hanna, *The Greenpeace Book of Coral Reefs*, New York, Sterling Publishing Co., Inc., 1992, p. 129.

15. Shinn, Eugene A., "What is really killing the reefs," *Sea Frontiers*, Vol. 35, No. 2, (March-April, 1989): 72-81.

16. Caroline S. Rogers, Interviews, 9/23/94, 1/96.

17. Gilberto Cintron-Molero, Interview, 7/12/94; Cintron-Molero, Gilberto, Ramon Martinez, Barbara S. Cintron, "Final Report: Effects of the M.V. A. Regina Grounding Mona Island, Puerto Rico," Puerto Rico Department of Natural Resources, Sept. 30, 1987.

18. Shinn, 72; Goenega, Carlos, *The State of Puerto Rican Corals: An Aid to Managers*, Mayaguez, Puerto Rico, University of Puerto Rico, May 1991, p. 24.

19. Aileen Velazco-Dominguez, Interview, 9/21/94.

20. Caroline S. Rogers, Interview, 11/2/94.

21. Aileen Velazco-Dominguez, Interview, 9/21/94.

22. Evelyn Wilcox, Interview, 9/1/94.
23. Hughes, 1550.
24. Ginsburg, Robert N. and Peter W. Glynn, "Summary of the Colloquium and Forum on Global Aspects of Coral Reefs: Health, Hazards and History," Miami, Rosenstiel School of Marine and Atmospheric Science, June 10, 11, 1993, p. v.
25. Goenega, 25.
26. Shinn, 81.
27. Smith and Buddemeier, 110.
28. Caroline S. Rogers, Interview, 11/2/94.
29. Wilkinson and Buddemeier, ix.
30. Hulm and Pernetta, 3.
31. Hulm and Pernetta, 13.
32. Wilkinson and Buddemeier, 6.
33. Hulm and Pernetta, 3.

CHAPTER 12 - OIL: THE UNWANTED SUBSTANCE IN OUR SEAS

1. American Petroleum Institute, "Facts About Oil," Washington, D.C., n.d.
2. Bordeaux, Christopher B., "Strategic Shipping Lanes," Energy Information Administration, *Petroleum Supply Monthly*, January 1994: xiii-xxi.
3. GESAMP Reports and Studies No. 50 "Impact of Oil and Related Chemical Wastes on the Marine Environment, IMO/FAO/UNESCO/WMO/WHO/ IAEA/UN/UNEP Joint Group of Experts on the Scientific Aspects of Marine Pollution, London, U.K.,1993: 180pp.
4. Bordeaux, xiv.
5. Energy Information Administration, *Petroleum Supply Annual* 1993, Vol. 1, June 1994,, 55; Environmental News Briefing, February, 1995.
6. Bordeaux, xxi.
7. Bordeaux, xxi.
8. Sankovitch, Nina , "Safety at Bay: A Review of Oil Spill Prevention and Cleanup in U.S. Waters," Natural Resources Defense Council, December, 1992., p. 3.
9. National Research Council, *Tanker Spills: Prevention by Design*, National Academy Press, 1991, p. xvii.
10. USCG, "An Assessment of Towing Vessel Manning Requirements," Office Of Marine Safety, Security and Environmental Protection, Washington, D.C., July, 1994, 42 pp; Gilberto Cintron-Molero, reviewer comments, 8/20/96.
11. International Maritime Organization, "World Maritime Day 1994," London, U.K., *IMO News*, No. 3, 1994, p. v.
12. *IMO News*, No. 3, v.
13. *IMO News*, No. 3, v., vi; Horscotci, "Ships of shame, inquiry into ship safety," Australian Government Publishing Service, Canberra, Australia, 1992.
14. The Coastal Area Contigency Plan for Puerto Rico and U.S.V.I., 1994, The U.S. Coast Guard, San Juan, Puerto Rico, 1994.

15. U.S. Department of Commerce National Technical Information Service, "Oil Spill Case Histories 1967-1991 - Summaries of Significant U.S. and International Spills," Seattle, Washington, NOAA, Nov. 1992, pp. 212, 231, 354, 358.
16. EPA, n.d.
17. Carlos Padin, Interview, 6/2/94; Pedro Gelabert, Interview, 5/30/94.
18. Jerry Barnett, Interviews, 7/20/94, 9/26/94; International Maritime Organization, "Oil pollution declines as shipping measures take effect," *IMO News*, No. 3:1993, p. 1.
19. Daniel Hamson, Interview, 7/23/94.
20. International Tanker Owners Pollution Federation, Ltd., "Information," London, U.K., January 1994, p. 3.
21. International Oil Pollution Compensation Fund, "IOPC Fund Report on the Activities of The International Compensation Fund in the Calendar Year 1993," London, U.K., 1994, p. 22.
22. *Golob's Oil Pollution Bulletin*, "API report finds U.S. spillage nearly doubled in 1994," Vol. VIII, No. 14, 7/12/96.
23. GESAMP, p.30.
24. GESAMP, p. 35.
25. Albers, Peter H., "Oil Spills and Living Organisms," Texas Ag. Ext. Serv., Texas A & M University, Publ. #B-5030, 1992: 1-9; GESAMP, p. 18; Gilbert Cintron-Molero, Interview, 9/13/94.
26. Aldeberto Bosque, Interview, 5/31/94.
27. Albers, 1992, 1; GESAMP, 41-43; Gilberto Cintron-Molero, Interview, 9/13/94.
28. Suchanek, Thomas H., "Oil Impacts on Marine Invertebrate Populations and Communities," *American Zool.*, 33(1993): 510-523.
29. ITOPF, "Response to Marine Oil Spills," London, U.K., 1986.
30. GESAMP, pp. 42, 43
31. Lee, R.F., "Processes affecting the fate of oil in the sea," Chapter 12 in Marine Environmental Pollution, 1. Hydrocarbons. Richard A. Geyer (ed.) Elsevier Oceanography Series, Elsevier Scientific Publishing Company, New York, 1980.
32. Albers, 1; Taft, Dr. William, "Brief Comments on the Fate and Effect of Oil Spills in the Marine Environment," Sarasota, Florida, Mote Marine Laboratory, n.d."; Suchanek, 511; GESAMP, 41-42; Gilberto Cintron-Molero, reviewer comments, 8/20/96.

CHAPTER 13 - HOW OIL AFFECTS THE MARINE WORLD

1. Gilberto Cintron-Molero, Reviewer comments, 8/21/96.
2. Albers, Peter H., "Oil Spills and Living Organisms," Texas Ag. Ext. Serv., Texas A & M University, Publ. B-5030, College Station, Texas, 1992, 9 pp.
3. Taft, Dr. William, "Brief Comments on the Fate and Effect of Oil Spills in the

Marine Environment," Sarasota, Florida, Mote Marine Laboratory, n.d., 3 pp.

4. Australian Department of Transportation, "Great Barrier Reef and Torres Strait Shipping Study, Vol. II," December 1994: 4-5, 4-14, 4-15, 4-22, 4-23, 4-28.

5. Burns, Kathyrn A., Stephen D. Garrity and Sally C. Levings, "How Many Years Until Mangrove Ecosystems Recover from Catastrophic Oil Spills?" *Marine Pollution Bulletin*, Vol. 26, No 5, 1993: 239-248.

6. Felix Lopez, Interview, 6/27/94; LCDR Sharp, Interview, 6/23/94.

7. Gilberto Cintron-Molero, Reviewer comments, 8/21/96.

8. Antonio Mignucci-Giannoni, Interview, 6/94.

9. Jerry O'Neal, Interview, 7/13/94.

10. Carlos Padin, Interview, 6/2/94.

11. James Timber, Interview, 6/22/94.

12. Gilberto Cintron-Molero, Interview, 9/13/94.

13. Aileen Velazco-Dominguez, Interview, 9/21/94.

14. Vance Vicente, Interview, 6/30/94.

15. Suchanek, Thomas H., "Oil Impacts on Marine Invertebrate Populations and Communities," *Amer. Zool.* 35:510-523 (1993); Dr. William Taft, "Brief Comments on the Fate and Effects of Oil Spills in the Marine Environment," Sarasota, Florida, Mote Marine Laboratory, n.d., 3 pp.

16. Goenega, Carlos *"The State of Puerto Rican Corals: An Aid to Managers,"* Mayaguez, Puerto Rico, University of Puerto Rico, May 1991, 56pp.

17. Kaplan, Eugene H. *A Field Guide to Coral Reefs Caribbean and Florida*, Boston, Peterson Field Guide Series, 1982, pp. 101-103.

18. Eugene A. Shinn, Interview, 6/24/94.

19. Dustan, Phillip, Barbara H. Lidz and Eugene A. Shinn, "Impact of Exploratory Wells, Offshore Florida: A Biological Assessment," *Bulletin of Marine Science*, 48(1): 94-124, 1991.

20. Wyers, S.C., H.R. Frith, R.E. Dodge, S.R. Smith, A. H. Knapp, and T.D. Sleeter, "Behavioral Effects of Chemically Dispersed Oil and Subsequent Recovery in Diploria Strisosa (DANA), P.S.Z. No I: *Marine Ecology*, 1986, 7(1): 23-42.

21. Wyers, *et. al.*, 37.

22. RPI International, Inc., "Tropical oil pollution investigations in coastal systems (Tropics) - final report; T.G.Ballou, R.E. Dodge, S.C. Hess, A.H. Knap, and T. D. Sleeter, RPI/R/86-19: Columbia, S.C., 1987, 223 pp.

23. Guzman, Hector M., Jeremy B.C. Jackson and Ernesto Weil, "Short-term ecological consequences of a major oil spill on Panamanian subtidal reef corals," Coral Reefs (1991) 10:1-12.

24. Guzman, 1991, 1.

25. Pennisi, Elizabeth "Blackened Mangrove, Smothered Reefs," *Science News*, Vol. 145, No. 15, (April 9, 1994): 232-233.

26. Pennisi, 233.

27. Burns, *et. al*, 245.

28. Burns, *et. al.*, 247.

29. Pennisi, 233.

30. Burns, et.al, 240.

31. Keller, B.D. and J.B.C. Jackson, "Longterm Assessment of the Oil Spill at Bahia Las Minas, Panama," synthesis Report, Vol. 1, Executive Summary, OCS Study, MMS 93-0047, U.S. Dept. of Interior, Mineral Management MGT Service, Gulf of Mexico OCS Region, New Orleans, LA, 129 pp, p. 56.

32. Ladner, Cornell M., James S. Franks, Jerry O'Neal, "Mississippi Coastal Waters Mineral Lease Sale Area Number 1 Environmental Profile and Generic Environmental Guidelines for Activities Associated with Oil and Gas Drilling Rigs and Production Platforms," Long Beach, MS., Department of Wildlife, Conservation Bureau of Marine Resources, January 1982, III-12, III-13; Suchanek, 519-521.

33. Taft, 3; Stickel, Lucille F. and Michael P. Dieter, "Ecological and Physiological/ Toxicological Effects of Petroleum on Aquatic Birds," Laurel, MD., U.S. Department of the Interior, Fish and Wildlife Service, July 1979: 1-13.

34. Albers, 3,4.

35. Ladner, III-12.

36. Ladner, III-13.

37. Stickel and Dieter, p. 7.

38. Taft, 3.

39. Suchanek, 515.

40. Rick Dawson, Interview, 10/94.

41. Sankovitch, Nina "Safety at Bay," (New York, Natural Resources Defense Council, December 1992): 63 pp.

42. GESAMP, "Impact of Oil and Related Chemicals and Wastes on the Marine Environment," IMO/FAO/UNESCO/WMO/WHO/IAEA/UN/UNEP (Joint Group of Experts on the Scientific Aspects of Marine Pollution), Rep. Stud. GESAMP, 1993: 180pp.

43. Evelyn Wilcox, Interview, 9/14/94.

CHAPTER 14 - HOW CORAL REEFS BENEFIT MAN

1. Hulm, Peter and John Pernetta, "Reefs at Risk - Coral reefs, human use and global climate change," IUCN, October, 1993: 24 pp.

2. Goenega, Carlos *The State of Puerto Rican Corals: An Aid to Managers*, Mayaguez, University of Puerto Rico, May 1991,: 56 pp.

3. Ward, Fred, "Florida's Coral Reefs are Imperiled," *National Geographic*, July 1990: p. 132.

4. Goenega, 8; Hulm and Pernetta, 21.

5. Hulm and Pernetta, 4.

6. Thorne-Miller, B. and Catena, *The Living Ocean, Understanding and Protecting*

Marine Diversity, Washington, D.C., The Island Press, 1991.
7. Gilberto Cintron-Molero, Interview, 9/13/94.
8. Tucker, Jonathan B., "Drugs from the sea spark renewed interest," *BioScience* Vol. 35, No. 9 (October 1985): 541-545.
9. Scheuer, Paul J., "Marine Resources: The Search for New Chemicals from Marine Organisms, *Sea Grant Quarterly*, Vol. 8, No. 4, Winter 1986: 1-8.
10. Tucker, 544; Livermore, Beth "Fishing For Cures," *Popular Science*, May, 1995: 62-64.
11. Kreeger, Karen Young "Researchers Plumb Depths to Fight a Wide Array of Human Diseases," *The Scientist*, Sept. 19, 1994.
12. Cheevers, Jack "Bacteria Gaining on Treatments," *The Denver Post*, 4/9/95.
13. Kreeger, 9/19/94.
14. Gilberto Cintron-Molero, Interview, 9/14/94.

CHAPTER 16 - RESPONSE MANAGEMENT AND LESSONS LEARNED

1. Robert Ross, Capt., Comments, "Spill Response Management and the *Morris J.Berman* Spill: The FOSC's Perspective," 1995 International Oil Spill Conference, Long Beach, California, 3/2/95.
2. IMO/IPIECA, "Special Session: International Report from: Arabian Gulf, Mediterranean, Caribbean, Southeast Asia, Latin America," 1995 International Oil Spill Conference, Long Beach, California, 2/28/95.
3. *Oil Spill Intelligence Report*, "To Improve Oil Spill Planning and Response Worldwide," Vol. XVIII, No. 30, August 17, 1995.
4. *Golob's Oil Pollution Bulletin*, "IMO Assembly adopts tighter budget for 1996 and 1997," Vol. VII, No. 25, December 1, 1995.
5. Walker, Ann H., D. L. Ducey, Jr., S.J. Lacey, and J.R. Harrald, "Implementing an Effective Response Management System," A White Paper Prepared for the 1995 International Oil Spill Conference, American Petroleum Institute, December 1994: 105 pp.
6. Robert Ross, Capt., Interview, 2/28/95.
7. Robert Ross, Capt., reviewer comments, 8/96; Robert Ross, Capt., Interview, 2/28/95; Dan Hamson, Interview, 7/23/94.
8. Jacqueline Michel, Reviewer Comment, 8/15/96.
9. Walker, et.al, pp. 42-45.
10. J.A. Nichols, Comments, White Paper Panel Session, "Establishing and Maintaining Response Capabilities," 1995 International Oil Spill Conference, Long Beach, California, 3/2/95.
11. Jean Francois Levy, Comments, White Paper Panel Session, "Establishing and Maintaining Response Capabilities," 1995 International Oil Spill Conference, Long Beach, California, 3/2/95.
12. John Schrinner, Interview, 8/15/94.
13. Dyer, Capt., USCG., Comments, White Paper Panel Session, "Establishing and Maintaining Response Capabilities," 1995 International Oil Spill

Conference, Long Beach, California, 3/2/95.

14. Walker, et.al., 12/94.

15. Vlaun, Richard C., Capt., Comments, "Preparedness Plans Versus Actual Response," 1995 International Oil Spill Conference, Long Beach, California, 2/28/95.

16. Edwin Stanton, CDR., Interview, 9/27/94.

17. Robert Ross, Capt., Interviews, 2/1/95, 3/1/95, 8/27/96, and reviewer comments 8/27/96.

18. Walker, et.al., pp. 47-48.

19. Ross, Robert, Capt., "The Response to the T/B *Morris J. Berman* Major Oil Spill," USCG, San Juan, Puerto Rico, 8/25/95, Table 5.

20. Etkin, Dagmar Schmidt, Ph.D., *The Financial Costs of Oil Spills*, Cuutter Information Corp., Arlington, MA., 1994, 254 pp., at pp. 51 and 52.

21. Ross, Robert, Capt., Interview, 8/19/96.

22. Jacqueline Michel, Ph.D., Reviewer comment, 8/15/96.

23. J. Card, Rear Admiral, USCG, Comments, Opening Session, 1995 International Oil Spill Conference, Long Beach, California, 2/27/95.

CHAPTER 17 - WHO PAYS WHEN OIL SPILLS?
THE INTERNATIONAL SCHEME

1. Sheehan, Daniel F., "OPA 90 and the International Regimes Concerning Oil Pollution Liability and Compensation: Are They So Far Apart, Must They Remain So." 1995 International Oil Spill Conference, Long Beach, California, 2/95; *Golob's Oil Pollution Bulletin*, "ITOPF and Cristal vote to end voluntary compensation schemes, Vol. VII, No. 25, 12/1/95; Jerry Barnett, Interview, 9/26/94; J. Nichols, Interview, 8/16/94.

2. Schrinner, J.E., "Overview of International Oil Pollution Conventions," presented to The Marine Emergency Management Course, Kingston, Jamaica, March 1992: 23 pp.; Sheehan, 2/95; International Oil Pollution Compensation Fund, "General Information on Liability and Compensation for Oil Pollution Damage," London, U.K., July, 1994; J. Nichols, Interview, 8/16/94; Hilary Rubin, Interview, 8/16/94.

3. IOPC, July, 1994; J. Nichols, Interview, 8/16/94.

4. IOPC, July, 1994; Hilary Rubin, Interview, 8/16/94; J. Nichols, Interview, 8/16/94; *Golob's Oil Pollution Bulletin*, "ITOPF and Cristal vote to end voluntary compensation schemes," Vol. VII, No. 24, 12/1/95.

5. IOPC, July, 1994; International Oil Pollution Compensation Fund, "Report Of the Seventh Intersessional Working Group: Fund/WGR.7/21," London, U.K., June 20, 1994; International Tanker Owners Pollution Federation Limited, "Note 1, Criteria for the Admissibility of Claims for Compensation, Preventative Measures," Fund/WGR.7/9/1, London, U.K., January 10, 1994.

6. ITOPF, Note 1, January 10, 1994.

7. ITOPF, Note 1, January 10, 1994:
8. ITOPF, Note 1, January 10, 1994.
9. Jerry Barnett, Interview, 9/26/94.
10. Jerry Barnett, Interview, 9/26/94.
11. Commonwealth of Puerto Rico, Department of Interior, U.S. Dept. Of Congress, NOAA, "Preassessment Screen Document, *Morris J. Berman* Oil Spill," San Juan, Puerto Rico, March 24, 1995.
12. Dr. Mark Miller, Interview, 5/13/96.
13. International Tanker Owners Pollution Federation, "Criteria For The Admissibility of Claims for Compensation, Note 2. Economic Loss," Fund/WGR.7/9/2, London, U.K., 1/10/94.
14. ITOPF, Note 2, 1/10/94.
15. ITOPF, Note 2, 1/10/94.
16. International Tanker Owners Pollution Federation Limited, "Criteria for the Admissibility of Claims for Compensation, Note 3. Environmental Damage," WGR.7/9/3, London, U.K., 1/10/94.
17. ITOPF, Note 3, 1/10/94.
18. ITOPF, Note 3, 1/10/94.
19. J. Nichols, Interview, 8/16/94; ITOPF Report, 6/20/94.
20. Hilary Rubin, Interview, 8/16/94.

CHAPTER 18 - WHO PAYS WHEN OIL SPILLS? THE U.S. SCHEME

1. National Pollution Funds Center, "Claimant's Information Guide," USCG, Arlingtion, Virginia, September, 1993; Sheehan, Daniel F., "OPA 90 and The International Regimes Concerning Oil Pollution Liability and Compensation: Are They So Far Apart, Must They Remain So?" 1995 International Oil Spill Conference, Long Beach, California, 2/95.
2. NPFC, September, 1993; Sheehan, 2/95.
3. J. Nichols, Interview, 8/16/94; J. Nichols, Comments, Session on White Paper, "Implementing an Effective Response Management System," 1995 International Oil Spill Conference, Long Beach, California, 3/2/95.
4. John Schrinner, Interview, 8/16/94.
5. Dan Hamson, interview, 10/94.
6. NPFC, September, 1993.
7. Comptroller General of the United States, "Decision in the Matter of U.S. Coast Guard - Oil Spill Liability Trust Fund, File: B-255979," Washington, D.C., October 30, 1995.
8. NPFC, September, 1993; Sheehan, 2/95; Donald Calkins, Interview, 8/1/94.
9. USCG, "News Release, Release #69-94, Claims," San Juan, Puerto Rico, 11/4/94.
10. Marisol Luna, Interview, 10/94.
11. *Golob's Oil Pollution Bulletin*, "Jury convicts parties involved in *Morris J. Berman* incident," Vol. VIII, No. 9. 5/3/96.

CHAPTER 19 - THE COST OF RECOVERY

1. J. Nichols, Interview, 8/16/94.
2. 61 Fed. Reg. 440, January 5, 1996.
3. 61 Fed. Reg. 440, January 5, 1996.
4. Hall, Douglas K., "Testimony Before the U.S. Committee on Environment and Public Works," Washington, D.C., 4/23/96.
5. Farron, Michael J., of the Confederated Tribes of the Umatilla Indian Reservation, "Testimony before the U.S. Senate Committee on Environment and Public Works, Washington, D.C., 4/24/96.
6. OPA 90, Subsection 1002(b)(2)(A); Office of General Counsel, National Oceanic and Atmospheric Administration, "Natural Resource Damage Assessment Rule - Oil Pollution Act of 1990," U.S. Dept. of Commerce, Washington, D.C., April, 1996.
7. 61 Fed. Reg. 440, January 5, 1996; *Oil Spill Intelligence Report*, "NOAA Issues New NRDA Proposal Substantially Changing 1994 Version," Vol. XVIII, No. 29, 8/10/95; *Golob's Oil Pollution Bulletin*, "House holds oversight hearing on NRDA programs," Vol. VII, No. 14, 7/14/95; *Golob's Oil Pollution Bulletin*, "NOAA restructures NRDA rule to stress resource restoration," Vol. VII, No. 17, 8/11/95; *Golob's Oil Pollution Bulletin*, "NOAA adopts restoration based NRDA process in final rule," Vol. VIII, No. 1, 1/12/96; NOAA, Office of General Counsel, "Natural Resources Damage Assessment Rule," U.S. Dept. of Commerce, April, 1996.
8. 59 Federal Register 1062, 1067, January 7, 1994.
9. Bennett, James F., B. Peacock and T. Goodspeed, "Computer Models for Damage Assessment Estimates of Use and Results," DOI and NOAA, Washington, D.C., February 27, 1995.
10. Barbier, E.B., M.C. Acreman and D. Knowler, "Economic Valuation of Wetlands: A Guide for Policy Makers and Planners," Ramsar Convention, Gland, Switzerland, 1996.
11. 58 Fed. Reg. 4601, 4610, January 15, 1993.
12. NOAA, "Natural Resource Restoration," U.S. Dept. of Commerce, Silver Spring, MD., November, 1995.
13. NOAA, "Reversing the Tide: Restoring the Nation's Coastal and Marine Natural Resources," U.S. Dept. of Commerce, Silver Spring, MD., November, 1995.
14. J. Nichols, Interview, 8/10/94.
15. *Golob's Oil Pollution Bulletin*, "NOAA adopts restoration-based NRDA process in final rule," Vol. VIII, No. 1, January 12, 1996.
16. *General Electric Company, et.al., Petitioners, v. United States Dept. of Commerce, NOAA, et.al., Respondents*, No. 96-1096 (and consolidated cases), U.S. Court of Appeals, (D.C. Cir.), "Petitions for Review," and "Motions for Leave to Intervene," 1996.
17. NOAA, "NOAA Responds to Challenges To Final Rule Under the Oil

Pollution Act of 1990," U.S. Dept. of Commerce, Washington, D.C., 4/15/96.

18. GE Suit, "Motion to Defer the Promulgation of A Briefing Schedule," May 8, 1996.

19. Mauseth, Gary S. and D.A. Kane, "The Use and Misuse of Science in Natural Resource Damage Assessment," A White Paper Prepared for the 1995 International Oil Spill Conference, American Petroleum Institute, Washington, D.C., December, 1994: 63 pp., p. 24.

20. Mauseth and Kane, December, 1994, p. 24.

21. Tom Campbell, Interview, 7/26/94.

22. J. Nichols, Interview, 8/16/94.

23. White, Dr. I.C., "Environmental Damage in the Context of the International Compensation Regimes," proceedings of ITOPF Seminar on NOAA Proposed NRDA Regulations, U.K., May 1994; Clark, A.F. Bessemer, "The Insurance Aspects of the Proposed NRDA Regulations," proceedings of ITOPF Seminar on NOAA Proposed NRDA Regulations and Their Implications for Vessel Interests, West of England P & I Club, May 9, 1994; International Tanker Owners Federation Ltd., "Criteria for the Admissibility of Claims for Compensation - Note 2. Economic Loss," IOPC Fund Seventh Intersessional Working Group, London, U.K., January 10, 1994.

24. Clark, p. 3, May 9, 1994.

25. James Bennett, Interview, 9/13/94; James Bennett, Reviewer Comments, 7/17/96.

26. White, May 1994.

27. *Golob's Oil Pollution Bulletin*, "House holds oversight hearing on NRDA programs," Vol. VII, No. 14, 7/14/95.

28. *Golob's Oil Pollution Bulletin*, "House holds oversight hearing on NRDA programs," Vol. VII, No. 14, 7/14/95.

29. H.R. 2500, November 21, 1995; Hall, Douglas K., "Testimony before the U.S. Senate Committee on Environment and Public works," Washington, D.C., 4/23/96.

30. J. Nichols, interview, 8/16/94.

31. U.S. Dept. of Transportation, *National Pollution Fund User Reference Guide*, Washington, D.C., May 1994; Sheehan, Daniel F., "OPA 90 and the International Regimes Concerning Oil Pollution, Liability and Compensation: Are They So Far Apart, Must They Remain So?" 1995 International Oil Spill Conference, Long Beach, California, 2/95.

32. Sheehan, 2/95.

33. Ted Armstrong, Interview, 10/4/94.

34. Ted Armstrong, Interview, 10/4/94.

35. *Golob's Oil Pollution Bulletin*, "USCG issues financial responsibility under OPA 90 and CERCLA," Vol. VIII, No. 6, 3/22/96.

CHAPTER 20: PREVENTION THROUGH PEOPLE

1. Card, J.C., "Prevention Through People," 60 Fed. Reg. 3288, 1/13/95; U.S. Dept. of Transportation, "Towing Vessel Inspection Study," U.S. Coast Guard Merchant Vessel Inspection and Documentation Division, 1994.
2. Pete Popko, CDR, USCG, Interview, 2/16/95
3. Congressional Record, House Debate, H8263, November 9, 1989.
4. Congressional Record, House Debate, H6271, August 1, 1990.
5. *Oil Spill Intelligence Report,* "USCG Issues Final Rule on Operational Measures for Single-Hull Tankers," Vol. XIX, No. 30, August 8, 1996.
6. *Golob's Oil Pollution Bulletin,* "Senate Bill Aims to Improve Spill Prevention and Response," Vol. VIII, No. 10, 5/17/96; S. 1730, June 20, 1996; Senate, Committee on Environment and Public Works, "Report together with Additional Views," Report 104-292, June 26, 1996.
7. International Maritime Organization, "1992 MARPOL amendments come into effect," *IMO News,* No. 2:1995.
8. International Maritime Organization, "International Maritime Prize for 1993," *IMO News,* No.1:1995.
9. Card, J.C., "Prevention Through People," 60 Fed. Reg. 3288, 1/13/95.
10. U.S. Dept. of Transportation, 1994.
11. Card, 60 Fed. Reg. 3288, 3289.
12. Card, 60 Fed. Reg. 3288, 3289.
13. Card, 60 Fed. Reg. 3288, 3289.
14. *Commonwealth of Puerto Rico vs. The M/V Emily S et.al,* U.S. District Court, District of Puerto Rico, 94-1019, "Order for Issuance of Warrant for Arrest," 1/8/94.
15. International Maritime Organization, "World Maritime Day 1993," *IMO News,* 3:1993.
16. Robert Ross, Capt., Interview, 2/28/95.
17. The Focus Group, *Licensing 2000 and Beyond,* USCG, Office of Marine Safety, Security and Environmental Protection, Washington, D.C., November, 1993.
18. Scott Glover, CDR, Interview, 2/14/95.
19. Scott Glover, CDR, Interview, 2/14/95.
20. Scott Glover, CDR, Interview, 2/14/95.
21. Robert Ross, Capt., Interview, 2/28/95.
22. International Maritime Organization, "World Maritime Day 1995," *IMO News* 3:1995.
23. International Maritime Organization, "The STCW Convention," *IMO News,* 2:1995; International Maritime Organization, "Draft STCW amendments ready for conference," *IMO News,* 1:1995.
24. Moore, William H. and K.H. Roberts, "Safety Management for the Maritime Industry: The International Safety Management Code," 1995 International Oil Spill Conference, Long Beach, California, February, 1995.

25. Moore and Roberts, p. 305.
26. Moore and Roberts, pp. 307, 309; International Maritime Organization, "IMO Code guidelines are approved,", *IMO News*, 2:1995; LCDR Sharp, Interview, 6/23/94.
27. LCDR Sharp, Interview, 6/23/94.
28. David A.Davidson, Comments, 1995 International Oil Spill Conference, Long Beach, California, 2/95.
29. Phillips 66, Lafayette Region Employees Newsletter, Vol. 5, No. 1, January-March, 1995.
30. International Maritime Organization, "World Maritime Day 1995," *IMO News*, 3:1995.
31. Oil Companies International Marine Forum, "Tanker Safety and Pollution Prevention: The OCIMF View on the Issues," London, September, 1993; Oil Companies International Marine Forum, "SIRE," London, n.d.; J.A. Nichols, Interview, 8/16/94.
32. *IMO News*, 3:1995.
33. *Golob's Oil Pollution Bulletin*, "IACS members agree to implement initiatives to improve vessel safety," Vol. VII, No. 13, June 30, 1995.
34. Caribbean Cooperation, "Report on the Second Preparatory Meeting on Port State Control," Port of Spain, October, 1994.
35. International Maritime Organization, "Marpol Convention is amended," *IMO News*, 1:1995; International Maritime Organization, "Port State control: towards global standardization," *IMO News*, 1:1994; Schrinner, John, "Overview of International Oil Pollution Conventions," presented to the Marine Emergency Mgt. Course, March, 1992.
36. Peter, Brian, "Safety Nets Protect Passenger Vessels," Proceedings of the Marine Safety Council, May-June, 1994.
37. Ron Northrup, Interview, 9/26/94; USCG, "Port State Control Initiative: Boarding Regime to Target Substandard Ships, Washington, D.C., 1994.
38. *IMO News*, No. 1:1995.
39. *IMO News*, No. 3:1995.
40. Ron Northrup, Interview, 9/26/94; USCG, "Port State Control Initiative," 1994.
41. Scott Glover, Interview, 2/14/95; Ron Northrup, Interview, 9/26/94; CDR Sharp, Interview, 6/23/94; Robert Ross, Capt., Interview, 2/28/95; Edwin Stanton, Interview, 9/27/94; Peter, May-June, 1994; USCG Port State Control Initiative, April, 1994.
42. Maria Pabon, Interview, 9/22/94; Sen. Freddie Valentin, Interview, 6/3/94; Joan Del Valle, Interviews, 6/3/94 and 9/21/94; Resolucion, R. del S. 492, 1/10/94; Ley, P. del S. 622, 3/9/94; Senado de Puerto Rico, "Informe Sobre La Resolucion Del Senado 492," 1994.
43. Robert Ross, Capt., Interview, 1/10/95.
44. Levy, Jean-Francois, "France and the Right of Intervention on the High Seas,"

1995 International Oil Spill Conference, Long Beach, California, 2/95; Jean-Francois Levy, Comments, in session of the 1995 International Oil Spill Conference, Long Beach, California, 2/28/95.

CHAPTER 21 - PROGRESS IN THE FIELD

1. U.S. Department of State, "The U.S. Coral Reef Initiative: Forging Partnerships for Effective Management," 12 pp., October 20, 1994.
2. Wilkinson, Clive R. and R.W. Buddemeier, "Global Climate Change and Coral Reefs: Implications for People and Reefs," UNEP-IOC-ASPEI-IUCN Global Task Force on Coral Reefs, 1994.
3. Wilcox, Evelyn, "Lessons from the Field: Marine Integration, Conservation and Development," World Wildlife Fund, 1994.
4. Lynn Davidson, Interview, 10/12/94; U.S. Department of State, "The International Coral Reef Initiative Working Paper," Washington, D.C., October 20, 1995.
5. Ginsburg, Robert N. and P.W. Glynn, "Draft Report, Results and Recommendations of the Coloquium on Global Aspects of Coral Reefs," Health, Hazards, and History, Rosenstiel School of Marine and Atmospheric Science, June, 1993.
6. Wilcox, Evelyn, 1994.
7. Ginsburg and Glynn, June, 1993.
8. U.S. Department of State, "The International Coral Reef Initiative: The Framework for Action, 1995.
9. Elizabeth Soderstrom, Interview, 3/26/96; Department of State, "The International Coral Reef Initiative: The Call to Action," 1995.
10. U.S. Department of State, "The International Coral Reef Initiative, Progress Report," January 9, 1995.
11. Pernetta, John C. (Comp.), "Monitoring Coral Reefs for Global Change," A Marine Conservation and Development Report, IUCN, Gland, Switzerland, 1993; Hulm, Peter and J. Pernetta, "Reefs at Risk - Coral Reefs, Human Use and Global Climate Change," IUCN, Gland, Switzerland, October, 1993.
12. Rogers, Caroline S., G. Garrison, R. Brober, Z. Hillis, and M. Franke, "Coral Reef Monitoring Manual for the Caribbean and Western Atlantic," National Park Service, Virgin Islands National Park, June, 1994.
13. IUCN, "Monitoring Coral Reefs," 1993.
14. Elizabeth Soderstrom, Interview, 3/26/96.
15. Elizabeth Soderstrom, Interview, 3/26/96.
16. Wirth, Tim, Under Secretary of the U.S., "New Global Reef Partnership Supports Biodiversity Goals," addressing the First Conference of the Parties, Convention on Biological Diversity, Nassau, the Bahamas, December 7, 1994.
17. Barzetti, Valerie, (ed.) "Parks and Progress: Protected Areas and Economic

Development in Latin America and the Caribbean," IUCN, Washington, D.C., 1993; Kelleher, Graeme and Richard Kenchington, "Guidelines for Establishing Marine Protected Areas," A Marine Conservation and Development Report, IUCN, Gland Switzerland, 1991; Wells, Sue and Nick Hanna, *"The Greenpeace Book of Coral Reefs,"* New York, Sterling Publishing Co., Inc., 1992.

18. Commonwealth of Puerto Rico, Law 150; Peter Ortiz, Interviews, 9/20/94, 10/11/94.

19. IUCN, Guidelines for Marine Protected Areas, 1991; Wilcox, Evelyn, "WWF Marine Conservation Initiative in the Wider Caribbean Region, Update," WWF, May, 1994.

20. IUCN, "Appendix 2, Resolution by 17th General Assembly of IUCN," 1988.

21. Barzetti, 1993; Kelleher et.al., 1991; Wells and Hanna, 1992.

22. Oceania Day, "Presentation: Paradise Under Pressure," Sixth Conference of the Contracting Parties, Ramsar, Brisbane, Australia, March 1996.

23. Evelyn S. Wilcox, Interview, 9/1/94; WWF, "1991 Annual Report," 1991; Wilcox, "Lessons from the Field," 1994; Barzetti, 1993.

24. Stolzenburg, "The Old Men & The Sea," *Nature Conservancy,* Nov/Dec. 1994; U.S. Agency for International Development, "Fragile Ring of Life" video, Washington, D.C., 1995.

25. Ward, Fred, "Coral Reefs are Imperiled," *National Geographic,* July 1990; Hirshfield, "Draft Management Plan Released for the Florida Keys Sanctuary," *Marine Conservation News,* Summer 1995.

26. Ward, p. 132.

27. USAID, 1995.

28. Ramsar, "Recommendation 6.7," Sixth Conference of Contracting Parties, Brisbane, Australia, March, 1996.

29. Gilberto Cintron-Molero, Interview, 9/13/94.

30. Gilberto Cintron-Molero, Interview, 9/13/94.

CHAPTER 22 - THE GREAT BARRIER REEF MODEL

1. Hulm, Peter and J. Pernetta, "Reefs at Risk - Coral reefs, human use and global climate change," IUCN, October, 1993; Ginsburg, Robert N. and P.W. Glynn, "Draft Report, Results and Recommendations of the Colloquium on Global Aspects of Coral Reefs: Health, Hazards and History," Rosenstiel School of Marine and Atmospheric Science, June,1993.

2. Gilberto Cintron-Molero, Interview, 9/13/94; John Passalacqua, Interview, 6/3/94.

3. Goenega, Carlos, *The State of Puerto Rican Corals: An Aid to Managers,* Department of Biology, University of Puerto Rico, Mayaguez, May, 1991.

4. Aileen Velazco-Dominguez, Interview, 9/21/94; *La Nueva Dia, Pesca Recreativa,* June, 1994.

5. Jorge Saliva, Interview, 7/11/94.
6. Susan Soltero, Interview, 6/1/94.
7. Wilcox, Evelyn, *Lessons From the Field: Marine Integration, Conservation, and Development*, World Wildlife Fund, 1994.
8. Jose Liboy-Gonzalez, Interview, 9/20/94; Boris Oxman, Interview, 9/20/94; Departamento de Recursos Naturales, "Guia Informativa Boyas de Anclaje," Puerto Rico, June, 1992.
9. Craik, Wendy, "Protecting the Great Barrier Reef from an Oil Spill," Great Barrier Reef Marine Park Authority, 1995 International Oil Spill Conference,1995; IUCN, "The Great Barrier Reef: A 25 Year Strategic Plan for the Great Barrier Reef World Heritage Area 1994 - 2019," April, 1994; Australia Department of Transportation, "Great Barrier Reef and Torres Strait Shipping Study," March, 1995.
10. Marine Environment Protection Services, "Protecting Our Seas," Australian Maritime Safety Authority, n.d.
11. Ottesen, Peter, S. Sparkes and C. Trinder, "Shipping Threats and Protection of the Great Barrier Reef Marine Park - The Role of the Particularly Sensitive Sea Area Concept," *The International Journal of Marine and Coastal Law*, Vol 9, No. 4, 1994; Australia Dept. Of Transportation, March, 1995.
12. Australia Bureau of Transport and Communications, "Major Marine Oil Spills, Risk, and Response," 1990.
13. Andrew Steven, Interview, 3/12/96.
14. Ottesen, et al., 1994; Marine Environment Protection Services, n.d.
15. Ottesen, et. al., 1994; Australia Dept. of Transportation, March, 1995; Australian Maritime Safety Authority, n.d.; Stephen Sparkes, Interview, 3/12/96; Jamie Storrie, Interview, 3/12/96.
16. IUCN, "Great Barrier Reef Strategic Plan," April, 1994; Andrew Steven, Interview, 3/12/96.
17. Australia Dept. of Transportation, March, 1995; Marine Environment Protection Services, n.d.; Marine Environment Protection Services, "National Plan," Australian Maritime Safety Authority, n.d.; Jamie Storrie, Interview, 3/12/96.
18. Andrew Steven, Interview, 3/12/96.
19. Jacqueline Michel, Ph.D., Reviewer comments, August 15, 1996.
20. Rogers, Caroline, "Efforts to Balance Marine-based Tourism with Protection of Coral Reefs and Seagrass Beds in a National Park," Coastal Resources Conference, University of Rhode Island, 1991.

CHAPTER 23 - WHAT *YOU* CAN DO

1. Rick Dawson, Interview, 10/94.
2. U.S. Environmental Protection Agency, "Coastal Non-Point Source Pollution," (Washington, D.C., n.d.)
3. U.S. EPA, n.d.

4. U.S. Fish and Wildlife Service, "Let's Clean Up Our Act! Critical Environmental Issues in the U.S.V.I.," (St. Thomas, V.I., Department of Planning and Resources, n.d.)
5. Barzetti, Valerie (ed.), "Parks and Programs, IUCN, 1993.
6. Terrascope, "Coral Reefs Face Precarious Future," *Our Planet*, Vol. 5, No. 3, 1993: 18.
7. Jose Berrios, Interview, 9/20/94; Berrios, Jose, Ramon Martinez, and Craig Lilyestrom, "Annual Progress Report for Project F-36 "Artificial Reef Research and Deployment," San Juan, Puerto Rico, Department of Natural and Environmental Resources Area of Living Resources Fisheries Division, September, 1994.
8. James Timber, Interview, 9/20/94; James H. Timber, Letter comments, 7/19/96.
9. Jose Berrios, Interview, 9/20/94.
9. Bohnsack, J. A., D.L. Johnson, R.F. Ambrose, *Ecology of Artificial Reef Habitats and Fishes*, (in Artificial Habitats for Marine and Freshwater Fisheries, Academic Press, Inc., 1991): 61-107.
10. Wilkinson Clive R. and R.W. Buddemeier, "Global Climate Change and Coral Reefs: Implications for People and Reefs," UNEP-IOC-ASPEI-IUCN Global Task Force on Coral Reefs, 1994.
11. Congressional Record, "House Debate on Amendment to OPA 90," H8263, November 9, 1989.

APPENDIXES

APPENDIX A

THE PR/USVI AREA CONTINGENCY PLAN
UNIFIED COMMAND SYSTEM ORGANIZATION

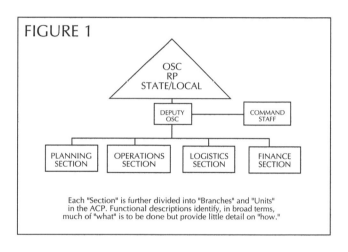

FIGURE 1

Each "Section" is further divided into "Branches" and "Units" in the ACP. Functional descriptions identify, in broad terms, much of "what" is to be done but provide little detail on "how."

MORRIS J. BERMAN
RESPONSE MANAGEMENT ORGANIZATION DAYS 2-7

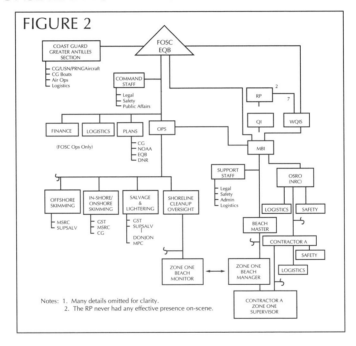

FIGURE 2

Notes: 1. Many details omitted for clarity.
2. The RP never had any effective presence on-scene.

MORRIS J. BERMAN
RESPONSE MANAGEMENT ORGANIZATION DAYS 8-42

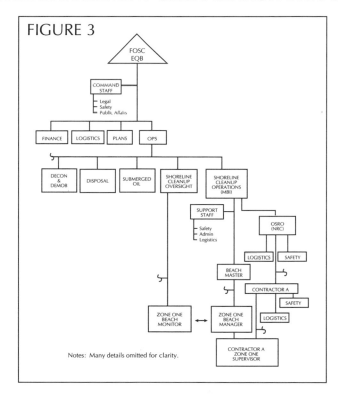

FIGURE 3

Notes: Many details omitted for clarity.

(Source: "The Response to the T/B *Morris J. Berman* Major Oil Spill," San Juan, Puerto Rico, USCG. Repreinted by permission of U.S. Coast Guard MSO, San Juan.)

APPENDIX B

NATIONAL RESPONSE CORPORATION
Berman Oil Spill Response San Juan, Puerto Rico
EQUIPMENT LIST

1- Marco Class XIC Belt Skimming System w/Guzzler B Vacuum Transfer Unit, 200 feet of Containment Systems 43-Inch Ocean Sweep Boom, and ACME Weir Skimmer Head (Owned by NRC)
1- Vikoma Cascade Weir Skimming Systems (Owned by NRC)
1- NRC-Dedicated Vikoma Cascade Weir Skimming System (Owned by Crowley Environmental Services)
1- Containment Systems 4-Band Rope Mop Skimming System (Owned by NRC)
1- FMT Weir Disk Skimming System w/Dedicated FMT Trailer (Owned by NRC)
1- NRC-Dedicated Desmi Ocean Weir Skimming System (Owned by Crowley Environmental Services)
1- NRC-Dedicated Crucial Rope Mop Skimming System (Owned by Crowley Environmental Services)
1- NRC-Dedicated Komara Disk Skimming Systems (Owned by Crowley Marine Services)
1- Sea Fab Portable Barge Set w/Guzzler A Vacuum Transfer Unit and ACME Weir Skimmer Head (Owned by NRC)
1- Kvichak 28-Foot Work Boat w/Dedicated Clemar Trailer (Owned by NRC)
12,000 feet of Containment Systems 21-Inch Inshore Boom (Owned by NRC)
5,000 feet of Oil Stop 43-Inch Ocean Boom (Owned by NRC)
4,000 feet of Oil Stop 43-1nch Ocean Boom (Owned by Crowley Environmental Services)
2,000 feet of American Marine 27-Inch Ocean Boom (Owned by Crowley Environmental Services)
6,000 feet of American Marine 12-Inch Inshore Boom (Owned by Crowley Environmental Services)
NRC-Dedicated 35,000-BBL Barge B251 (Owned by Crowley Environmental Services)
Crowley Marine Services 35,000-BBL Barge B250-4

SUPSALV, USN: *Berman* Oil Spill Response
San Juan, Puerto Rico
EQUIPMENT LIST

SYSTEM NO.	DESCRIPTION	QTY
SK0711	Class V Oil Recovery Vessel	3
SK0712	Skimmer Rack	3
SK0924	Skimmer, Oil, Sorbent Belt, Class XI (VOSS)	1
PU0851	Heavy Oil Transfer System	1
WB0722	Boom Handling Boat, Rigid, 24 ft.	6
OB0800	Oil Storage Bladder, 136K, Type L	3
OB0809	Oil Storage Bladder, 26K, Type E	3
P08100	Floating Hose System	1
WB0717	Inflatable Boat, 15 ft.	1
AC0330	Air Compressor	4
GE0404	Diesel Generator Set	1
PW0045	Hydraulic Power Unit, Mod 6	3
HC0003	Hydraulic Hose Reel	1
VA0727	Command Van	1
VA0010	Rigging Van	1
VA0508	Workshop Van	1
VA2119	Cleaning Van	1

(Source: Naval Sea Systems Command, "Final Report *Morris J. Berman* Spill San Juan, Puerto Rico," 7 January - 1 February 1994, 7/26/94. Reprinted by permission of U.S. Coast Guard MSO San Juan.)

APPENDIX C

ACRONYM LIST

ACEC	Assessment, Communication, Education and Commitment
ACP	Area Contingency Plan
AMSA	Australian Maritime Safety Authority
BFI	Browning Ferris Industries
BGI/BGPR	Bunker Group/Bunker Group of Puerto Rico, Inc.
CARICOMP	Caribbean Coastal Marine Productivity Program
CERCLA	Comprehensive Environmental Response, Compensation and Liability Act
CFCs	Chlorofluorocarbons
CFR	Code of Federal Regulation
CITES	Convention on International Trade in Endangered Species of Wild Flora and Fauna
CLC	International Convention on Civil Liability for Oil Pollution Damage
COI	Certificate of Inspection
COFR	Certificate of Financial Responsibility
CRRT	Caribbean Regional Response Team
CSN	Caribbean Stranding Network
CV	Contingent Valuation
CVM	Contingent Valuation Method
CWA	Clean Water Act
DEC	Department of Environmental Conservation
DOI	U.S. Department of the Interior
DNR	Department of Natural and Environmental Resources

DOS	U.S. Department of State
DOT	U.S. Department of Transportation
EPA	U.S. Environmental Protection Agency
EQB	Environmental Quality Board (Puerto Rico)
FDA	U.S. Food and Drug Administration
FOSC	Federal On-Scene Coordinator
FWPCA	Federal Water Pollution Control Act
FUND	International Convention on the Establishment of an International Fund for Compensation for Oil Pollution Damage
GANTSEC	Coast Guard Greater Antilles Section
GBR	Great Barrier Reef
GBRMPA	Great Barrier Reef Management Park Authority
GESAMP	Advisory body of experts to United Nations
GST	Gulf Strike Team
HAZMAT	Hazardous Materials
HRMT	Heritage Resource Management Team
IACS	International Association of Classification Societies
ICRI	International Coral Reef Initiative
ICS	Incident Command System
IMO	International Maritime Organization
IOPC	International Oil Pollution Compensation Fund
IPC	Institute of Puerto Rican Culture
IPIECA	International Petroleum Industry Environmental Conservation Association
ISM	International Safety Management Code
ISPR	Incident Specific Preparedness Review
ITOPF	International Tanker Owners Pollution Federation Ltd.

IUCN	World Conservation Union
LAPIO	Low API Gravity Oil
MARPOL	International Convention for the Prevention of Pollution from Ships
MBI	Maritime Bureau, Incorporated
MSIS	Marine Safety Information System
MSO	U.S. Coast Guard Marine Safety Office
MSRC	Marine Spill Response Corporation
NCP	National Contingency Plan
NGO	Nongovernmental Organization
NOAA	National Oceanic and Atmospheric Administration
NPFC	National Pollution Funds Center
NPS	National Park Service
NRC	National Response Corporation
NRDA	Natural Resource Damage Assessment
NRDC	Natural Resources Defense Council
NRT	National Response Team
NSF	National Strike Force
OCIMF	Oil Companies International Marine Forum
OPA 90	Oil Pollution Act of 1990
OPRC	International Convention on Oil Pollution Preparedness Response and Co-Operation of 1990
OSC	On-Scene Coordinator (same as FOSC)
OSLTF	Oil Spill Liability Trust Fund
OSIR	Oil Spill Intelligence Report
OSRO	Oil Spill Removal Organization
OSSM	On-Scene Spill Model
P & I	Prevention and Indemnity Clubs

PSSA	Particularly Sensitive Sea Area
QI	Qualified Individual
RCP	Regional Contingency Plan
RMS	Response Management System
RP	Responsible Party
RRT	Regional Response Team
SAIT	Shoreline Assessment/Inspection Team
SARA	Superfund Amendments and Reauthorization Act
SHPO	State Historic Preservation Office
SIRE	Ship Inspection Report Programme
SMT	Spill Management Team
SOLAS	International Convention for the Safety of Life at Sea, 1974
SONS	Spill of National Significance
SSC	Scientific Support Coordinator
STCW	Standards of Training, Certification, and Watchkeeping for Seafarers Convention, 1978
SUPSALV	U.S. Navy Supervisor of Salvage
TAD	Temporary Additional Duty
T/B	Tank Barge
UCS	Unified Command Structure or System
UN	United Nations
UNCLOS	1982 United Nations Convention on the Law of the Sea
UNEP	United Nations Environment Program
UNESCO	United Nations Educational, Scientific and Cultural Organisation
USAID	U.S. Agency for International Development
USCG	U.S. Coast Guard

USFWS	U.S. Fish and Wildlife Service
USGS	U.S. Geological Survey
USN	U.S. Navy
USVI	U.S. Virgin Islands
UTV	Uninspected Towing Vessels
VOSS	Vessel of Opportunity Skimming System
WHC	1972 Convention on the Protection of the World Cultural and Natural Heritage
WQIS	Water Quality Insurance Syndicate
WWF	World Wildlife Fund

APPENDIX D

INTERVIEW SUBJECTS

Aledo, Freddie, Ranger, National Park Service, San Juan, Puerto Rico, 9/20/94.

Armstrong, Ted, Assistant Division Chief, Vessel Certification, National Pollution Funds Center, 10/4/94.

Baca, Bart, Ph.D., 7/22/94.

Barker, LCDR., U.S. Coast Guard, Merchant Vessel Marine Inspection, Standards and Development Branch, 10/5/94.

Barnett, F.G. (Jerry), Regional Consultant on Marine Pollution - Wider Caribbean, International Maritime Organization, 6/1/94, 7/20/94, 9/26/94.

Benggio, Bradford, LCDR, Scientific Support Coordinator, NOAA Hazmat, 7/7/94.

Bennett, James F., Environmental Protection Specialist, U.S. Dept. of the Interior, 7/28/94, 9/13/94.

Berrios, Jose M., Natural Resources Specialist, Dept. of Natural Resources, Commonwealth of Puerto Rico, 9/20/94.

Boot, Richard, Attorney, U.S. Dept. of Justice, 7/28/94.

Bosque, Aldeberto, Environmental Engineer, U.S. Environmental Protection Agency, 5/31/94.

Burlington, Linda, Office of General Counsel, NOAA, 10/12/94.

Calanni, A. Carlo, S.E. Regional Response Manager, Marine Spill Response Corp., 7/96.

Calkins, Donald, Chief of Claims, National Pollution Fund Center, 8/1/94.

Campbell, Tom, Attorney, Entrix Inc., 7/26/94.

Chasis, Sarah, Attorney, Natural Resources Defense Council,10/4/94.

Cintron-Molero, Gilberto, Internal Affairs Specialist, USFWS, Dept. of the Interior, 7/12/94, 9/13/94.

Collins, Pat, Attorney, National Pollution Fund Center, 8/31/94.

Corrada, Rosa, Attorney, Office of the Attorney General, 9/28/94.

Csulak, Frank, Oceanographer, NOAA, Damage Assessment Center, 7/14/94.

Dawson, Rick, Marine Biologist, USFWS, 10/94.

Davidson, Lynn, Environmental Solutions International,10/12/94.

Del Valle, Joan, Assistant to Senator Valentin, Senate of Puerto Rico, 6/3/94, 9/21/94.

DeMarco, Greg, ICF Inc., 10/5/94, 4/95.

Dodge, Richard E., Ph.D., Associate Dean and Coral Reef Studies, Oceanographic Center, 7/22/94.

Drake, Susan, Bureau of Oceans and International Environmental and Scientific Affairs, Dept. of State, 10/19/94.

Eichenberg, Tim, Attorney, Center for Marine Conservation, 6/5/95.

Fratalone, Joseph, Assistant U.S. Attorney, U.S. Attorney General's Office, 7/11/94.

Garamone, Matthew, Environmental Scientist, U.S. EPA, 8/30/94, 3/1/95.

Garcia-Diaz, Jorge, Director, Office of Legal Affairs, Environmental Quality Board, Commonwealth of Puerto Rico, 8/23/94.

Garcia, Mayra, Department of Natural Resources, Commonwealth of Puerto Rico, 9/21/94.

Garcia, Neftali, Ph.D., Servicios Cientificos Y Tecnicos, 6/3/94, 10/27/94.

Gelabert, Pedro, Director, Dept. of Natural Resources, Commonwealth of Puerto Rico, 5/30/94.

Glover, Scott, CDR, Merchant Vessel Personnel, U.S. Coast Guard, 2/14/95.

Gonzalez, Blanche, Coordinator of Damage Assessment, Environmental Quality Board,   Commonwealth of Puerto Rico, 7/19/94, 9/20/94.

Good, Rick, Regional Director, BFI Waste Systems, Inc., 7/27/94.

Ferra, Lourdes, Sales Representative, BFI Waste Systems, Inc., 9/21/94.

Gonzalez, Fernando, Ph.D., Dept. of Chemistry, University of Puerto Rico, 10/4/94.

Hamson, Daniel, Chief of the Environmental Response, Planning and Assessment Dept., U.S. Park Service, 7/23/94.

Hankins, Paul, U.S. Naval Sea Systems Command, 10/11/94, 8/96.

Graves, Yuri, Ensign, U.S. Coast Guard, Pollution Response Branch, 9/14/94.

Hayes, Chris, Attorney, U.S. Coast Guard, Pollution Response Branch, 8/31/94.

Hess, John, Assistant Branch Chief of U.S. Coast Guard National Response Center, 8/30/94.

Hooper, Richard, Supervisor of Salvage and Diving, U.S. Navy Sea Systems Command, 10/14/94.

Hubbard, Jerry, Lt., U.S. Coast Guard, Gulf Strike Team, 8/2/94.

Jimenez, Christopher, Federal On-Scene Coordinator, U.S. EPA, Response and Prevention Branch, 7/29/94.

Johnston, James, Biologist, National Biological Survey, 8/23/94.

Lee, James, Regional Environmental Officer, U.S. Dept. of Interior, 6/30/94.

Liboy-Gonzalez, Jose, Director, Puerto Rico Coastal Zone Management Program, Commonwealth of Puerto Rico, 9/20/94.

Lopez, Felix, Field Contaminant Specialist, USFWS, 6/27/94.

Luna, Marisol, Catano, Puerto Rico, 10/14/94.

Martinez-Lorenzo, Patricio, Attorney, Commonwealth of Puerto Rico, 6/5/94.

McKinley, Audrey, LCDR, U.S. Coast Guard, 6/4/94, 7/20/94.

Michel, Jacqueline, Ph.D., Director, Environmental Technology Division, Research Planning, Inc., Columbia, South Carolina, 8/96.

Mignucci-Giannoni, Antonio, Marine Biologist, Scientific Coordinator for Red Caribeña de Varamientos, 6/94.

Miller, Mark, President, National Response Corp., 8/25/94, 8/26/94.

Miller, Mark, Dr., neurobiologist, Institute of Neurobiology, University of Puerto Rico, 10/8/94.

Morales, Deborah, The Tourism Company, 10/27/94.

Morton, Mary, Attorney, Office of Environmental Policy and Compliance, Dept. of the Interior, 7/28/94, 9/13/94.

Nichols, J.A., Technical Manager, International Tanker Owners Pollution Federation Ltd., 8/16/94.

Northrup, Ron, Lt., Inspector, U.S. Coast Guard, 9/26/94.

Oland, James, Field Supervisor of Caribbean Field Office, USFWS, 7/11/94.

O'Neal, Jerry, Toxicologist, USFWS, 7/13/94.

Ortiz-Rosas, Peter, Director, Division de Patrimonio Natural, Dept. of Natural Resources, Commonwealth of Puerto Rico, 9/20/94, 10/14/94.

Oxman, Boris L., Consultant, Department of Natural Resources, Commonwealth of Puerto Rico, 9/20/94.

Pabon, Maria, Attorney, Office of the Attorney General, 7/12/94, 9/28/94.

Padin, Carlos, Division of Coastal Zone Management, Commonwealth of Puerto Rico, Dept. of Natural Resources, 6/2/94.

Pantel, Agamemenon Gus, Ph.D., Anthropologist, San Juan, Puerto Rico, 8/2/94, 9/20/94.

Passalacqua, John De, Professor of Law, University of Puerto Rico, 6/3/94.

Pereira, Nelson, Petty Officer, Marine Environmental Response Branch, U.S. Coast Guard, 7/19/94.

Peters, Kevin, Miami Divers, 10/10/94.

Popko, Pete CDR, Assistant Division Chief, Merchant Vessel Inspection & Document Division, U.S. Coast Guard, 2/16/95.

Praschak, Andrew, Attorney, U.S. EPA, 6/30/94.

Ramos, Rosa, EPA Citizen Coordinator, 7/25/94.

Rodríquez, Abimael, Professor, Dept. of Chemistry, University of Puerto Rico, 10/11/94.

Rogers, Caroline, Biologist and Chief Scientist, National Biological Survey, U.S. Virgin Islands National Park, 9/23/94, 11/2/94, 1/4/95, 1/5/95.

Roque, Pablo, Lt., Inventory Officer, Marine Safety Office, U.S. Coast Guard, San Juan, Puerto Rico, 7/18/94.

Ross, Robert, Capt., former Commanding Officer U.S. Coast Guard Marine Safety Office, Captain of the Port, San Juan Puerto Rico, 1/10/95, 1/11/95, 2/1/95, 2/28/95, 8/96, 8/19/96.

Rubin, Hilary, International Oil Pollution Compensation Fund, 8/15/94.

Saliva, Jorge, Ph.D., Avian Ecologist, USFWS, 7/11/94.

Santiago, Carlos, Ph.D., Biochemist, University of Puerto Rico, 10/4/94.

Scannel, Cheryl, Attorney, NOAA Natural Resource Damage Assessment Center, 7/20/94.

Schafler, Jonathan, Chief Park Ranger, U.S. Park Service, San Juan, Puerto Rico, 7/12/94, 9/20/94.

Schrenk, William, Natural Resource Defense Council, 10/6/94, 10/12/94, 3/95.

Schrinner, John, Regional Consultant on Marine Pollution, International Mari-time Organization, 8/15/94.

Sharp, LCDR, U.S. Coast Guard, National Pollution Fund Center, Claims, 6/23/94.

Shinn, Eugene, Geologist, U.S. Geological Survey, Coastal Resources Program, 6/24/94.

Singhota, Gurpreet, Technical Officer, Marine Environment Division, International Maritime Organization, 8/15/94.

Soltero, Susan, News Broadcaster, *TeleOnce*, San Juan, Puerto Rico, 6/1/94.

Stanton, Edwin, CDR, U.S. Coast Guard, Marine Safety Office, San Juan, Puerto Rico, 5/31/94, 9/27/94, 3/1/95.

Timber, James, Biologist, Natural Resource Specialist II, Dept. of Natural Resources, Commonwealth of Puerto Rico, 6/22/94, 9/21/94, 8/96.

Torres, Genaro, Chairman, Environmental Quality Board, Commonwealth of Puerto Rico, 7/1/94, 9/28/94.

Velazco-Dominguez, Aileen, Marine Scientist, Coordinator of Aquatic Resources Education, Dept. of Natural Resources, Commonwealth of Puerto Rico, 9/21/94.

Valentin-Acevedo, Freddie, Senator, Senate of Puerto Rico, 6/3/94.

Vicente, Vance, Biologist-Oceanographer, Chief Scientist for NOAA in the Caribbean, 6/30/94.

Walker, Stewart, Coordinator, U.S. Coast Guard, 10/11/94, 2/14/95.

Webb, Richard, Oceanographer, Water Resource Division, USGS, 6/15/94.

Wilcox, Evelyn, Marine Coordinator, Latin American and Caribbean Program, World Wildlife Fund, 9/1/94.

INDEX

INDEX

A

ACEC, 175, 263
Agenda 21, 256
Aguadilla, 41, 55, 60, 65, 67, 74, 77, 89, 279
Aledo, Freddie, 65
alga
    Calcareous, 105
    Halimeda, 105
American Petroleum Institute, 133, 221
AMSA, 270, 273
area contingency plan (ACP), 20, 22-23, 43, 84, 170, 273
aromatics, 79, 142
artificial habitat ecology, 280
artificial reefs, 278-279, 281

B

Bahia Las Minas Spill, 140-145
baseline data, 144
Benggio, Brad, 57, 59, 61, 63, 66, 78
Bennett, James, 224
benthic, 47-48, 50, 142
Berman Enterprises, Inc., 5,6
Berman/Frank, 5-8, 15, 86, 173, 207, 227-228, 233, 245
Berrios, Jose, 279
BFI Landfill, 73-74
BGI Trader, 5-8, 28, 30-32, 37-38
bioactive molecule, 160-161
biological diversity, 152-153
biomedical model, 158
bioremediation, 74-75
Biotechnology, 153
black band disease, 114
bleaching, 116-117, 143

boom, 30, 32-37
Bunker Group of Puerto Rico, Inc., 5-6, 88, 211
Bursatella leachii (same as *Busatella*), 156-158, 201

C

Calanni, Carlos A., 41
Campbell, Tom, 223
capacity building, 254-255
Captain of the Port, 27, 89, 245
Card, Admiral James C., 191, 232
Caribbean Regional Response Team, 84
Caribbean Petroleum Corp., 6
Caribbean Responder, 34, 41, 44-45
Caribbean Stranding Network, 51-53
Caribe Hilton (see Hilton Hotel), 37, 49, 67, 72
CARICOMP, 257
CERCLA, 213, 225-226
Certificate of Financial Responsibility (COFR), 174, 194, 213, 225
Certificate of Inspection (COI), 7-8
Chafee, Sen. John H., 17, 209, 231
choke points, 127, 170, 263
Cintron-Molero, Gilbert, 122, 142, 152, 165, 261
CITES, 121
CLC, 132, 197, 204, 225, 269
Clean Water Act, 83, 207, 213
climate change, 97, 108, 116-117, 124-125, 149, 255
coastal zone, 101
Comptroller General, opinion, 194, 209-210

Condado, 120, 130-131, 162, 202
contingent valuation, 174, 218-219
coral
    coral polyp, 102-103
    stony coral, 102-105, 108, 116-117, 279-280
corallite, 103, 107
COREXIT 9580, 61-62, 82
CRISTAL, 197
Crowley Environmental Service of Puerto Rico, Inc., 30
Crown of Thorns starfish, 115

D
Dawson, Rick, 52-53, 148, 275
decant, 69, 72
Defense Dumping Grounds, 39-40
Dept. of Natural and Environmental Resources, 124
DESMI, 34
dispersants, 61, 198
Dodge, Richard E., Dr., 144
Dos Hermanos Bridge, 67
double hull, 230-231, 246
Doyle, Larry, Capt., 6
dredging, 69, 71-72, 74, 97, 124
Dustan, Phillip, 143

E
Eason, Tom, 68
Eason Diving and Marine Contractors, Inc., 68
echinoderms, 47, 53, 58, 146
Ecotourism, 252
El Cañuelo, 78
El Morro, 78
Emanuel, George, 11-13

Environmental Protection Agency, 83
Environmental Quality Board, 59, 84, 247
Escambrón, 47-49, 54, 61, 67, 74
Exxon Valdez, 52-53, 83, 99, 147-148, 169, 190-191, 205, 209, 230, 232

F
Federal on-scene coordinator, 32
fleshy macroalga, 114
Focus Group, 235
Fort San Geronimo, 82
Fund Convention, 132, 194-197, 204

G
Gaia Hypothesis, 152
Garcia, Mayra, 63, 222
General Electric Company, 221
GESAMP, 134, 136, 149
global warming, 97, 117, 255
Goenega, Carlos, 124, 151, 264, 266
Gonzalez, Fernando, Dr., 160-161
Good, Rick, 74, 76
gorgonian, 47-49, 105, 142, 159-160, 163, 280
Great Barrier Reef, 103, 106-107, 115, 256, 266-267
Great Barrier Reef Marine Park Authority, 266-268, 273
Greenhouse effect, 116
Gulf Strike Team, 30-32, 40
Guzman, Hector, 144, 278-280

Miller, Mark, NRC, 36-37, 156, 201
molecular compound, 161
monitoring, 149, 152, 255
mousse, 19, 35, 135, 137, 147
MSIS, 227, 243-244
mystery spills, 50-51, 65

Oil Spill Removal Organization,
22, 85
OPA 90, 24-25, 43-44, 83, 85-86,
133, 169-170, 174, 176-179, 181-
183, 193-194, 205-209, 211-213,
216-218, 221, 224-226, 228, 230-
232, 273, 278, 281
OPRC, 170, 178-179, 183, 269
option, 203

N

National Contingency Plan, 27, 84,
177, 182
National Historic Preservation
Act, 78
National Historic Register, 78
National Historic Sites, 78
National Oceanic and
Atmospheric Administration,
39, 209, 213
National Park Service, 22, 59, 64
National Research Council, 129
National Response Corp, 30, 36, 88
National Response Team, 84
National Strike Force, 31, 185, 188
Natural Resource Damage
Assessment, 174, 213-214, 222
Natural Resources Defense
Council, 129, 222, 224
New England Marine Services,
Inc., 5
No. 6 fuel oil, 3, 58-59, 78-79

P

Padin, Carlos, 142
Palau, 259-260
Panama Canal, 127, 130
Pantel, Agamemnon Gus, Dr., 78-
80
passing ship syndrome, 181-182
Pennekamp Park, 260
Pereira, Nelson, 65-66
Persian Gulf, 128
Peters, Kevin, 27-30, 38
petrochemicals, 98, 127
petrohighway, 128, 176
photosynthesis, 104, 117, 120
planula larva, 102, 106, 122, 124
Preassessment damage, 200
Protocols, 195-197, 202, 207, 238
PSSA, 269-270

O

O'Neal, Jerry, 141
O'Neil, William, 244
OCIMF, 229, 240-241
Oil Spill Liability Trust Fund
(OSLTF), 24, 88, 194, 209-211

Q

Qualified Individual, 85

R

Radisson Normandie, 67, 72
RAMSAR, 260
Rebuttable presumption, 221-222, 225
Red Caribeña de Varamientos, 51
REEFPLAN, 273
Regional Response Team, 84
response management system, 179, 183, 187, 190-191
Responsible Party, 21-22
Restoration, 200, 202-203, 215, 217-218
    compensatory, 217-218
    primary, 217
Rivera, Pedro, 6-7, 9-12, 14-15, 86, 88, 211
Rodríguez, Abimael D., Dr., 158, 161, 163
Rogers, Caroline, 113, 123, 273
Ross, Robert Capt., USCG, 13, 16, 21, 27, 41, 44-46, 83, 86-87, 89-93, 185-189, 190, 199-201, 248

S

Safety Management Certificate, 239
safety nets, 25, 172, 242
Saliva, Jorge, 53-55, 265
salvage, 4, 19, 27, 30-31, 38, 40-41, 43, 46, 85, 87, 92, 129, 183, 185, 193, 200
San Juan Harbor, 3, 14, 37-38, 58, 86, 89-91, 130-131, 228
Sand farms, 75
Santiago, Carlos, Dr., 154, 156, 162
seagrass, 59, 67, 69, 110-111, 139-140, 142, 144-146, 169
sedimentation, 97, 104-105, 124-125

sewage, 97, 115, 120, 243
shackle, 8, 12, 16, 64
Shaks Beach, 41, 55, 67, 89
Sharp, CDR, 240
sheen, 136
Shinn, Eugene, 107, 113, 125, 143-144
Shoreline Assessment/Inspection Team, 59, 63
single hull, 28, 230
skimmer
    Belt Incliner, 33-36
    DESMI, 34
    VOSS, 34, 45
    Weir, 33-34, 73
soft eye, 12
SOLAS, 171, 173, 231, 237-240, 244
Solomon Islands, 258
Soltero, Susan, 67, 265
sorbent snare, 61, 69
Spill of National Significance, 36, 84, 169, 187
Stanton, Edwin CDR, 21, 28, 30, 37, 41, 53, 58, 72, 171, 185
State Historic Preservation Office, 78
STCW, 171, 174, 229, 235-238
submerged oil, 35, 45, 67-69, 87, 199
Sun Oil Company, 32
Superfund Act, 224
symbiotic relationship, 104
system
    closed system, 181-182
    open system, 181, 183-184, 187

T

Taft, William, Dr., 140, 147
Tajamar Ruins, 77
tarballs, 58-59

tarmats, 47, 54-55, 59, 63
thimble, 9, 12-14, 16
Timber, James, 47-48, 142
Torres, Genaro, 64, 75, 267, 271, 273
TOVALOP, 196-197
Towing Vessel Safety Act, 173
towline, 3, 5, 8-14, 16, 207, 245
Trajectory modeling, 61
transcription factor, 155
trophic cascading, 148
Type A, 214, 217-219, 222

U

U.S. Dept. of the Interior, 222
U.S. Fish and Wildlife Service, 50, 260
U.S. Navy, 4, 30, 170
U.S. Virgin Islands, 84, 122
UNEP, 179, 257
UNESCO, 256-257, 271
Unified Command Structure, 22, 171
USAID, 252, 254-256

V

Valentin Acevedo, Freddie, Sen. , 246-248
Values
    Bequest, 203, 215-216
    Existence value, 215
    Heritage value, 203, 215
    Interim lost values, 215
    Option value, 215
Vicente, Vance, 49-50, 62, 142

W

waste disposal, 74
Water Quality Insurance Syndicate, 87, 221, 224
waterfall effect, 155
WHC, 260-261
White, I.C., Dr., 24, 224
white band disease, 114
Wilcox, Evelyn, 124, 149, 251, 265
Wilkinson, Clive 125

Z

Zooxanthella, 103, 117, 144